KU-206-855

TOYSHOP
STEAM

TOYSHOP STEAM

BASIL HARLEY

 Model and Allied Publications

Argus Books, 14 St. James Road,
Watford, Herts

First published 1978

© Basil Harley, 1978
© Argus Books Ltd., 1978

ISBN 0 85242 583 X

All rights reserved. No part of this book may be reproduced in any form without the written permission of Argus Books Ltd.

HAMPSHIRE COUNTY LIBRARY

621.1
0228

085242583X

C001868134

Printed and bound in Great Britain by
A. Wheaton & Co. Ltd., Exeter

ACKNOWLEDGEMENTS

I AM INDEBTED to many people for their help and encouragement in the preparation of this book, particularly to Eric Malins and his son Stephen for so cheerfully putting up with my many visits to their works in Brierley Hill and my innumerable questions. My friend Tom Nevins, Secretary of the Stuart International Model Engineers' Club, was largely responsible for the original concept of the book and has offered many helpful suggestions. I am also most grateful to M. B. Jenkins of Ridlington, Norfolk, for much of the information about his father (together with the photograph) which is incorporated in Chapter Two, a substantial part of which appeared in the *Model Engineer*. My thanks are due to Mr Martin Evans, former editor, for permission to reproduce it here.

Among those who have contributed much useful information are Roy Hallsworth, A. G. Wheeler, Jim Gamble and Robert Brown of the Museum of Childhood, Menai Bridge. My thanks are also due to Edward L. Simmons and A. Berg of the trade journal *Games and Toys* for so kindly putting their archives at my disposal and for allowing me to use some early advertisements as illustrations.

All the Mamod engines illustrated belong to Malins (Engineers) Ltd and I am indebted to them for the use of the photographs. I am grateful to Vic Smeed of *Model Boats* for letting me use the photograph of his Perico launch, to Roy Hallsworth for the Meccano steam engines and to the Jensen Manufacturing Co, U.S.A.

Acknowledgement is made to the *Birmingham Post* for photographs of Malins' works and to the Museum of Science and Industry, Birmingham for the picture of William Murdock's miniature locomotive.

The other toys are in my own collection and the colour photographs of them were taken by Graham P. Hughes of Birmingham.

v

før Jessie

CONTENTS

Half-title illustration: A portable engine with a chain drive incorporated to make it into a primitive traction engine. Probably made by Marklin in Germany about 1914. The boiler lagging has been added since.

Frontispiece: A typical toy workshop driven by an elegant mill engine by the German maker, Bing. The workshop was made about 1904 by Ernst Plank.

PREFACE

THE FLYING
DUTCHMAN STEAM LOCO-
MOTIVE. This is the cheapest real
steam working Locomotive ever
brought out, and is invented and
manufactured solely by us. Large
metal boiler with four wheels, oscillating cylinder, steam tap, safety
valve, steam pipe and furnace complete. Warranted to work. Carriage
free, 3s. 3d.   *c. 1886*

ALMOST UNTIL THE end of Queen Victoria's reign the technology of the industrial countries of the world depended upon steam power. The Industrial Revolution, fathered in Britain by such people as Abraham Darby, Thomas Newcomen and James Watt in the 18th century and George and Robert Stephenson in the early 19th, had changed the face of what had been mainly an agricultural country. As prosperity grew and the middle classes became more comfortable they were able to indulge their children with complex and advanced toys reflecting the new technological society.

So it was that in such countries as Britain, Germany and France there grew up a mechanical-toy industry of which the toy steam-engine, be it locomotive, ship, mill engine or generating station, was an important part. These toys had an enormous appeal to boys whose ambitions were already being expressed as 'when I grow up I want to be an engine driver'. In other countries where the steam-engine came much later there was no such tradition in the nursery.

Over the years toy steam-engines have had much competition for youngsters' attention but even when this has been at its most intense there have always been some manufacturers to cater for the faithful. One of these, Malins (Engineers) Ltd, has outlasted all its rivals in Britain and today is the sole maker here of these exciting 'toys for engineers', as Mr Eric Malins describes them. Nor is this in any way a tenuous survival, for Malins Ltd is a lively, thrusting family firm with a third-generation director on the board. It has a modern design and production programme fully in tune with the needs of the last quarter of the twentieth century.

Today there is a tremendous nostalgia for steam—be it on the railway or on the rally field among the traction-engines or, most satisfactorily, in the home. This book tells something about the development of the steam-toy business in Britain and in particular the story of over forty years' production of Mamod steam-engines.

ix

x

Chapter One

THE BIRTH
OF A TRADITION

A LOVE OF steam-engines has been a deeply-rooted part of our character for close on two hundred years and never has it been more widespread than now when fewer and fewer steam-engines—in factories, ships, on rail or on the roads—are in normal daily use. This appeal is understandably at its strongest in Britain where the first steam-powered machines in the world were developed at the very beginning of the 18th century and where steam made its greatest impact on our social life, in the reigns of George IV, William IV and the youthful Queen Victoria, through the development of the railways.

As far back as 1792 Dr Erasmus Darwin was much impressed by the power of steam and in his prophetic poem 'The Economy of Vegetation' he wrote:

> *Soon shall thy arm, unconquered Steam! afar*
> *Drag the slow barge, or drive the rapid car;*
> *Or on wide-waving wings expanded bear*
> *The flying chariot through the fields of air.*

Then there was Ebenezer Elliot who wrote in the poem 'Steam at Sheffield' in the 1830s:

> *Oh, there is glorious harmony in this*
> *Tempestuous music of the giant, Steam,*
> *Co-mingling growl, and roar, and stamp, and hiss,*
> *With flame and darkness! . . .*

And who does not remember 'MacAndrew's Hymn', Rudyard Kipling's celebration of the great marine engines of the late 19th century, or even the awful poems of William McGonagall such as 'Success to the Newport Railway' and 'The Tay Bridge Disaster'?

Alongside this adulation of the vast engines that powered the later phases of the Industrial Revolution there has always been the fascination of the miniature. And

1

since the doll's house was the natural miniature for the girls to play with so it was that the toy steam-engine was, and still is, its natural counterpart for boys. In the absence of the real thing a miniature version has great merit as a substitute, but 'miniature' can mean different things to different people. In the homes of the rich it could mean a 15″ gauge steam railway round the park, like that built by Sir Arthur Heywood at Duffield Bank in Derbyshire in 1874, or the public miniature railways such as the Ravenglass and Eskdale or the Romney, Hythe and Dymchurch opened in the 1920s.

But the majority of the proud possessors of miniature railways or steam yachts had to be content with much smaller things than that. From the middle of Victoria's reign specialist toy firms were including steam-engines in their catalogues and practically every optician of any standing stocked a brass locomotive or two, or at least a few tiny glass Aeolipiles—popular versions of Hero of Alexandria's pre-Christian reaction steam turbine. A typical early advertisement by George Richardson and Co of Liverpool in the *Boy's Own Book* of 1868 says that they

have published a catalogue of 100 pages illustrated with full page engravings of Steam Engines, lithographed drawings beautifully printed in colours, of Screw and Paddle-steamers, and sailing Yachts and views of magic lanterns and slides and full particulars of Microscopes, Telescopes etc.

And they would send copies free on receipt of a stamp!

The advertisement itself lists seven vertical steam-engines with oscillating cylinders, brass flywheels and safety-valves ranging in price from 6s (30p) to £1 11 6d (£1.60); two horizontal engines with fixed cylinders; two twin-cylinder marine engines at £3 10 0d (£3.50) and £5 5 0d (£5.25) and two locomotives. The specification for the locomotive costing £2 10 0d (£2.50) is as follows:

Locomotive Engine, with Boiler, brass Chimney, Safety Valve, two brass Cylinders, with Steam Boxes, Water Tap and Pipe, Tender, Spirit Lamp with two jets, mounted on six brass flanged wheels.

The larger one, costing £3 15 0d (£3.75) is simply described as

Ditto, Ditto, very superior.

Although the locomotive engine wheels were flanged it is very unlikely that they were intended to run on rails—these came much later. These early locomotives, bringing something of the magic of the Stephensons, George and Robert, into the nursery, were affectionately known as Dribblers. The exhaust steam and water from their oscillating-cylinder engines was not turned up the chimney but allowed to dribble out from small holes in the 'steam boxes' onto nursery or kitchen floors. It is worth mentioning too, that in these early days we are only talking about the locomotives; coaches and wagons were not made for such toy engines for the very good reason that they were not powerful enough to pull them along. Indeed, a short

2

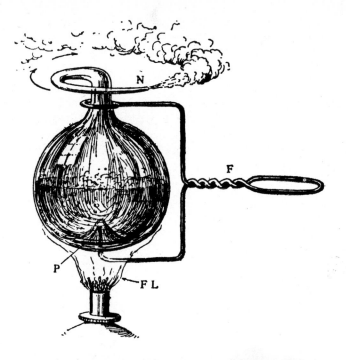

About A.D. 100 Hero of Alexandria first demonstrated a primitive reaction steam turbine. He called it an Aeolipile or Sphere of the Winds. Many versions have been made since. Here is a glass one as sold in opticians' shops about seventy years ago.

sputtering dash across the room was about all that a Dribbler, such as that shown in Plate 1, was capable of, and then only if the driver had had the forbearance not to sound the whistle first.

Richardson's of Liverpool were not the only toy merchants who sold steam-engines. There was for instance the very oldest of them all, John Bateman and Co of London calling itself the Original Model Dockyard, founded in 1774. Then there was the Clyde Model Dockyard established in 1789 and Steven's Model Dockyard started in 1845 (and which also advertised in the *Boy's Own Book*). At first these model dockyards provided ship models to the Admiralty, but by the 1880s they had moved into the toy field and were already covering the basic range of toy steam-engines which were to be the staple fare of little boys for the next thirty years or so.

First came the locomotives which took pride of place with most boys. Any who had travelled on, or even seen, an early train could not rest for long without longing to possess such a treasure in miniature. They were not all as expensive as Richardson's —for example Theobald's of London sold an engine *The Flying Dutchman* complete and warranted to work for 3s 3d (16½p) carriage paid.

Then there were the stationary engines reminiscent of those to be seen perhaps in the local sawmill or small engineering works, or even in the barn of a progressive farmer and landowner. And, to many of us the most satisfying of all, the toy

3

steamboats, either with screw-propellers or side paddle-wheels, which could make long mysterious trips across the open waters of a nearby lake, with all the excitement and hazards of a stalled engine or even potential shipwreck.

At the same time an increasing number of books for boys published drawings and instructions for making simple steam-engines at home. This was not the easiest thing to do without a lathe, even if some parts were bought ready-made from firms like Stevens. Nevertheless boys were cajoled and encouraged to attempt it by a number of writers, such as the Reverend J. Lukin, for instance, in the *Young Mechanic*, which was published in 1886. He even suggested that an attempt be made to make a model of Newcomén's engine and showed some inadequate drawings 'to help'.

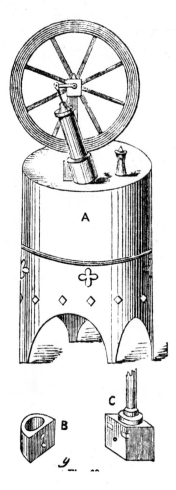

The Young Mechanic, an instruction book for boys published in 1886, showed how to make this typical 19th century toy steam-engine with its spirit-fired boiler and oscillating-cylinder engine mounted on the boiler top.

A number of suppliers either made, or had made for them, small, cheaply produced novelties such as simple steam-operated turbines, often only a brass fan spinning in a pin-hole steam jet from a spirit-fired pot boiler. Such 'engines' had

4

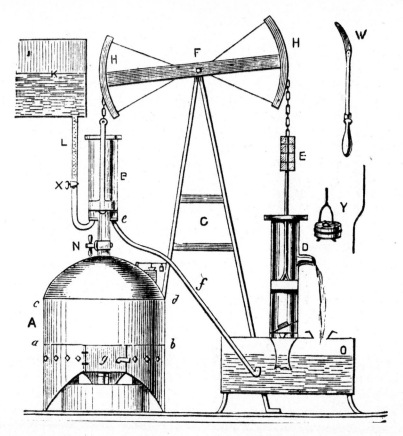

Also described in *The Young Mechanic* was this simplified version of Newcomen's atmospheric engine — though difficult to make and even more difficult to get to work properly.

little power but demonstrated a principle and were capable of driving tiny toys like this dancing figure sold in the 1870s.

Dancing Nigger by Steam (Patent) 2/6 [13½p]. Selling by thousands. This ingenious and laughable contrivance is very unique; the nigger dancer being worked by means of a novel steam engine constructed entirely of metal and with polished brass works, and when in motion the nigger dances in various graceful attitudes for one hour at each operation. Packed in boxes with directions and carriage paid for 40 stamps. From R. F. Benham manufacturer and patentee, office 1 Gresham Buildings, Guildhall EC. Descriptive catalogue embellished with illustrations of novelties, seven stamps.

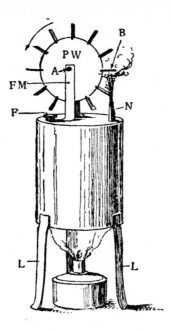

In 1629 Giovanni Branca described a primitive-impulse turbine and these, too, have been used as very simple toys. They attain high speeds but develop very little power.

However, the truly characteristic British steam-driven toy was solidly made of brass-turned parts, castings and solid-drawn tubing. As the 19th century drew to its close, however, competition from the tinplate toymakers of Germany became more and more intense. Most of them were centred in or around Nuremberg. Their products were attractive, gaudily painted and had a great appeal for children; above all, they were cheap. Since all the makers were primarily metalworkers with factories equipped with press tools, lathes, drilling machines and brass finishers' shops it was easy to include steam-engines alongside the increasingly popular clockwork trains and the newly-introduced electrically operated railway layouts.

It was not long before vast quantities of German toys were being imported into Edwardian England every year. Among those who included steam-engines in their catalogues were Gebr. Marklin; Gebr. Bing who owned the biggest toy manufactury in the world, employing over five thousand people by 1914; Ernst Plank; Georges Carette which was a French firm working in Nuremberg; J. Falk; J. Schoenner and many others. Then there were countless more who sold to wholesalers like Moses Kohnstam (Moko) and whose goods were merely marked 'Made in Germany' to satisfy import regulations — originally American. Some companies made specially for British retailers such as Gamages, but many British toy and model shops included German toys (and German-produced engravings of them) in their catalogues, without saying who made them or even that they were foreign. A good example of this can be seen in a 1905 catalogue of Jasper Redfern of Sheffield, which listed locomotives made by Bing and Carette and some splendidly ornate battleships by Ernst Plank without a word to say where they came from.

6

An interesting version of the popular electricity generating station by Bing of Nuremberg. This was introduced about 1908 and sold in this country by Bassett-Lowke. The engine is geared 4:1 to the Siemens-type dynamo.

Many of the designs of these toys had an air of gaiety, of fantasy even which, whilst clearly adding to their appeal to children (for whom, after all, they were intended) have for many of us today an almost unbearable nostalgia. It may be a steam fire-engine perhaps with a rubber hose for squirting water, attendant firemen and escape ladders, or a miniature generating station to make enough electricity to light a sister's doll's house.

However, such an influx of gaudy toys did not appeal to everyone and it must be admitted that often the products of the anonymous makers, and some others, did not last very long. In general, sturdiness was not a quality that the Germans sought to sell and many of the steam-engines were badly underpowered. Most of the railway engines bore scant resemblance to any known prototypes and the more fastidious boys were already expressing their discontent at what was available when Mr W. J. Bassett-Lowke came upon the scene.

The Lowkes had been engineers in Northampton for some years during the 19th century and young Wenman became a skilled model engineer during his apprenticeship with the family firm. He soon started a small business of his own supplying castings for the use of other model engineers. In 1900 he went to the great Paris Exhibition where the firm of Bing had a large stand. Here he first met Stephen Bing, the then head of the Nuremberg company. Two years later the first Bassett-Lowke catalogue included a large number of Bing and other German products. Among them were steam stationary engines, 'stork-leg' 2-2-0 locomotives, hot-air engines and some remarkably ugly traction-engines and steam-rollers.

This type of toy locomotive was a development of the dribblers and was often known on the Continent as a 'stork leg' engine.

Soon Bassett-Lowke had met and persuaded a locomotive designer, Henry Greenly, to design some really representative models of contemporary railway engines and rolling-stock which he got Bing (and later Georges Carette) to manufacture. So the *toyshop* railway got a real lift and in the hands of Bassett-Lowke was soon to blossom into the *model* railway. This was particularly true of the larger sizes and it was not long before the firm, with Henry Greenly as designer, was making steam passenger-carrying miniature railways for private estates and seaside pleasure grounds

Here perhaps is a point in our story at which we might pause to consider the difference between a toy engine and a model engine. The first distinction to recognize is that a toy is designed, made and sold primarily as a children's plaything. It may well be crude but it may also have great charm. Indeed its charm may well lie essentially in a comparatively primitive or untutored attempt to reproduce the real prototype in miniature. Also it is often made in an inappropriate material, such as a wooden train or a cardboard doll's house. Thus toys differ in their very nature, as well as intention, from the detailed scale model which is usually made in the same materials and in a similar manner to the original that it portrays. A model, however, can also be something made on a small scale by an engineer or a designer to test out a theory — to see if a proposed machine is likely to work — at of course a much lower cost than building a full-sized prototype. Such models were made by the earliest

8

engineers such as Thomas Newcomen, James Watt and William Murdock, whose original three-wheeled 'locomotive' is now in the Museum of Science and Industry, Birmingham.

William Murdock's miniature 'locomotive' made about 1781 or 1784 had its single cylinder set within the rectangular boiler. This is the original model. It is about 18″ long and can be seen in the Museum of Science and Industry, Birmingham.

There is naturally a grey area between the poor model and the good toy and, indeed, for the purpose of this book we shall not be considering W. J. Bassett-Lowke's engines in much greater detail because, though at their least expensive they represented the finest 'toys' available before the Great War, they were otherwise truly in the 'model' class.

Others in this grey area were those British-made engines appearing in the early Gamage's Christmas Catalogues. David and Charles have published a reproduction of the 1913 issue which contains some pages of solid 'High class English made steam engines' some of which, designed for pressures up to 100 p.s.i. and having bores and strokes of 1½″ are clearly not children's playthings. Others, however, are, but the word Gamage under an illustration is no guarantee that it is British, since the company bought in from many sources and put their own name on the goods.

A. BELL

24, Silk Street, Whitecross Street, London, E.C.1.

Call and see the following Range of Goods.

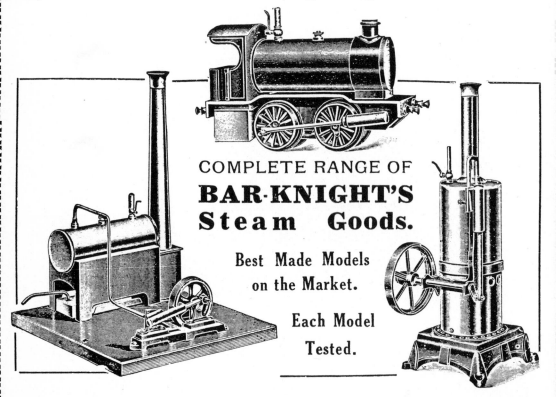

COMPLETE RANGE OF

BAR-KNIGHT'S
Steam Goods.

Best Made Models
on the Market.

Each Model
Tested.

MORGAN NATURE TOYS

BRAGG-SMITH'S PLANES

Full Range of

KENT TOY CO.'S

**Fruit, Vegetables, Cone Birds, etc., Picture
Studio Puzzles, Ogilvie's Furniture Sets, and**

MY OWN GOODS.

By the end of 1914 neither Britain nor Germany were making many toys since little could be spared from the war effort in either country for such luxuries. It was not until the beginning of the 1920s that production was resumed — often of old designs hopefully to be sold in a world that had changed irrevocably. But during the half century or so that had elapsed since the *Boy's Own Book* of 1868 there had been established a firm and lasting love of toy steam-engines among generations of budding engineers.

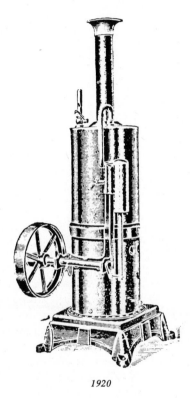

1920

Opposite: A. Bell sold, among other toys, a range of engines made in Glasgow by Bar-Knight and Co. Here is one of their advertisements from the trade journal *Games & Toys* of 1920.

Chapter Two

THE STEAM-TOYMAKER OF DEREHAM

AFTER THE UNPRECEDENTED slaughter of the Great War it took the western world some time to recover from both its social and its economic effects. Not surprisingly there remained an intensity of anti-German feeling in Britain that extended even into the 1930s. This, allied to a worsening economic climate at the beginning of the decade encouraged a fervent patriotism that looked askance at all foreign (particularly German) wares and extolled the virtues of solid, well-designed goods made by British craftsmen.

This was no less true of the toy trade than of other and perhaps more important industries. The Germans found the competition increasingly severe and because of this, and from other causes, Gebr. Bing went bankrupt early in the 1930s and ceased to manufacture. At about the same time Bond's o' Euston Road, an old-established toy and model company, had a grand sale of some £3,000 worth of surplus Marklin stock at prices which make many of us today weep for such lost opportunities. British toys for British boys had become an effective slogan.

As far as toy steam-engines were concerned perhaps the single most effective exponent of this philosophy was Mr Geoffrey Bowman Jenkins whose Bowman range of engines, launches and locomotives are remembered with great affection by many who spent their childhood in the 1920s and 1930s.

Bowman Models as a company started about 1923. It was established and run by Mr Jenkins in association with Hobbies Ltd of Dereham in Norfolk. Mr Jenkins, who was born on November 24th 1891, was an ingenious, inventive man and his first patent concerned a toy locomotive. The patent (No 123464) was granted in 1919 and covered a means of driving a toy railway engine by elastic. His second patent (No 208319) related to 'Improvements in Toy Steam Boats and the Like' and was granted in 1923. At the time of the application he was making toy boats in London near to Clapham Junction. As a result of his success and with the patent in his pocket, so to speak, he was invited to Norfolk to discuss a joint venture.

Hobbies Ltd had been going since 1897 and was well established in the fretwork and amateur woodworking business. They had a number of retail shops in such cities as Birmingham, Manchester, Glasgow and Leeds, as well as three in London. Their

Geoffrey Bowman Jenkins in later life — about 1953 when he was in his
early 60s.

factory in Dereham was fully equipped with a woodworking shop, machine shop,
press shops and a complete iron foundry. In addition to the woodworking and
fretwork designs, for which they were so well known, they also manufactured their
own extensive range of fretsaws, lathes and other woodworking machinery. They
were the obvious people to make the wooden hulls (and clearly their marketing
strength was also significant) and Mr Jenkins, as Bowman Models, would make the
engines. This he did in premises, near to the railway station, rented to him by
Hobbies. At first about a dozen men were employed but later this trebled as the
business developed.

The patent had established the form of almost all the English toy steam launches
in the 1920s and 30s. They were totally different in appearance to any of the
elaborate German boats; they also had wooden hulls and much more powerful
engines. In their simple clean lines they looked very much like pre-1914 racing
launches with a high length-to-beam ratio, sloping counter and turtle-back spray
hood.

The patent was not, however, a very strong one and soon a number of other
manufacturers had paid them the compliment of making very close copies. The
Hobbies-Bowman range soon extended to four — birds of somewhat dissimilar

13

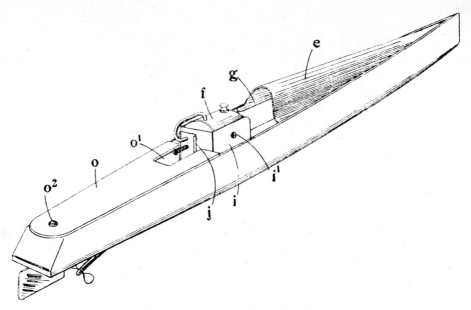

The Bowman Jenkins Patent of 1923 (No 208,319) included this drawing of his proposed steamboat.

Here is an early launch in exactly the same style and proportions as the patent drawing. Part of the engine support is made of wood.

feather — *Swallow* (20″ long), *Snipe* (23″ long), *Eagle* (28″ long), and *Seahawk* (also 28″ long). All had powerful single-acting oscillating-cylinder engines (*Seahawk* was twin-cylindered) with ample boiler capacity and windproof casings and lamps. They were tested — the prototypes only in all probability — on a large water-tank on the

14

premises, and the late Mr R. Pratt, one-time deputy chairman of Hobbies Ltd, recollected at least one early model exploding dramatically in the middle of a run.

Boats such as these, and *Miss America, Peggy* and a simple steam tug became very popular in the late 1920s and the *Meccano Magazine* and the *Boy's Own Paper* carried numerous advertisements for them. Few were as exciting as that for the *Snipe*. One of these—perhaps the one shown on page 18—was so effective that it resulted in an enquiry, from Lagos as it was remembered, asking how far it would travel in the open sea and, please, how many adult passengers will it carry! One boys' hobby annual published elaborate (but incomplete) instructions for making your own, flagrantly copying the Bowman design. The article started 'Model power boats are rather expensive things to buy, but it is not at all difficult to make one . . .' Bowman's ranged from 17s 6d (88½p) to £2.12s (£2.10)—trifling by today's standards.

Peggy and Miss America were introduced in 1923. This advertisement of 1928 shows improved versions, but the prices remained the same. Peggy was still selling at 22s 6d (£1.12½) in 1938.

It was not long before the range of products was extended to include stationary engines and locomotives. These were not all made jointly with Hobbies Ltd, but some were sold under Bowman Models' own name and others under Hobbies' name, or even the names of other members of the trade. For instance, in 1927 Warboys and Smart of Dereham were advertising engines 'made under Jenkins patents' and, one

15

BRITISH

MADE

·REGD·
TRADE MARK

Power!!
Lifts 25 lbs.!

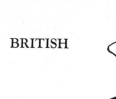

Are you tired of toy steam engines? Then listen! Here at last is an engine with real power behind it, one which will actually lift 25 lbs. as illustrated. And yet the price is no more than the better type foreign toy engine which will not lift a few pounds. It is our last word after 10 years' concentration on steam engine design and mass production methods. Compare its massive proportions and solid British workmanship with other engines at £3, £4 and £5, and you will realize what an astounding advance has been made.

Don't buy a toy engine this Xmas

Buy a Bowman Model E101. Use it as a toy while you will and then you can settle down to serious model engineering, knowing that your engine will fulfil every need. Our illustration gives no idea of its size and solidity. Note the dimensions in specification and if you want further conviction ask your dealer to get one on appro. for you to see.

SPECIAL ADVANTAGES

Runs for 1½ hours.
2,000 R.P.M. guaranteed.
Continuous drip feed oiling.
2 driving pulleys, for high and low gear.

If unable to obtain, we will gladly send direct. If not absolutely satisfied we return your money. Our good name is behind this engine—we guarantee it unconditionally.

┌─────────────────────────────┐
│ **A FEW STOCKISTS** │
│ **All Hobbies Branches,** │
│ **Hamleys,** │
│ **Civil Service Stores,** │
│ **Lorberg, Wiles, Kelly.** │
└─────────────────────────────┘

BOWMAN MODELS
(Dept. 26)
DEREHAM, Norfolk

16

25 LBS

SPECIFICATION

ENGINE.
Our special design. Brass piston made steam tight by special process. Mounted on solid steel frame. Heavy flywheel of solid turned brass, 3" by ¾", weighing over 1 lb. 7 to 1 reduction gear of brass. Both axles of ¼" steel rod with driving pulley on each. Ends turned down to take Meccano size pulleys, etc. Drip feed oiling ensuring efficiency and long life. Steam exhausts through funnel.

BOILER.
Solid drawn, highly polished brass, 6"×3" with safety valve and whistle, mounted on firebox of steel, painted with special heat proof paint.

LAMP.
Safety type for methylated spirit. Made of polished brass.

BASE.
Hardwood, 11"×9", painted with heat proof paint.

PRICE

Complete with filler and instructions in strong box

32'6

Post U.K. 1/3 extra.

Post abroad usually 6/-, any excess returned.

MODEL E 101

BRITISH MADE

Wormar Steam Engines

NEW SUPERHEATED MODELS

Specially designed for Meccano, with standard drilled base (by kind permission of Messrs. Meccano Ltd.) the new improved Wormar engines offer the following advantages which are shared completely by no others, *regardless of price.*

1. Meccano drilling. The only engine really suitable for building into any Meccano model.
2. The only seamless *brazed* boiler at the price. The usual soldered boilers leak if water boils dry.
3. Separate engine with solid steel frame—detachable steam union. Can be taken to pieces and re-assembled or run some distance from boiler (by adding extra steam tubing).
4. Superheated steam giving great power and speed. Double power of foreign engines and more than most British ones, even at higher prices.
5. Models E & C will take Meccano chain drive.

UNIQUE TO WORMAR USERS

This drip feed oil adaptor is another unique advantage. You can add it to your engine at any time. Secured by one bolt only and ensures maximum power and wearing qualities.

All small steam engines lose half their power if not constantly oiled. For Models D & E only, but handy workers can adapt to Mod. C.

Drip feed Oiler, Price 1/-, Post U.K. 1½d.

Wormar special oil for above, 1/- tin, post 2d.

EXPERIENCE COUNTS

We have concentrated on Meccano steam engines. We know your needs. This unapproachable 5/11 model is the result.

The ELITE
Model E. Single Cylinder, with geared drive and nickel plated base.

The famous TROJAN. Model D. Single Cylinder with filler, etc.

5/11

Post U.K. 7d. extra.

10/6

Post U.K. 9d. Oiler fitted as illustrated 1/- extra.

INTERESTING BOOKLET FREE.

The SUPER
Model C. Twin Cylinder. with geared drive and nickel plated base.

15/6

Post U.K. 9d.

Postage abroad is usually 2/3¡ either model. Any excess returned.

Of all Meccano Dealers or

Worboys & Smart, 189, Station Yard, Dereham

Opposite: By 1927 Bowman Jenkins was taking full page advertisements in such magazines as *The Boy's Own Paper* and the *Meccano Magazine.*

Captain Smart of Warboys and Smart was a cousin of Bowman Jenkins who in fact made these engines. An advertisement from 1927 or thereabouts.

17

STEAM ALONE CAN GIVE sustained SPEED

A 'Snipe' hits up racing pace from the start—and speeds on and on for mile-long cruises! How is it done?

What secrets of design enable the 'Snipe' to overtake and out-distance every boat in its class? First, *live* steam. Secondly, the Bowman *double-power* engine which gets the very last ounce out of every pound pressure. And finally the 'Snipe's' wood hull which is scientifically stream-lined to cut through the water. The 'Snipe' is a real speedboat in miniature (length 23 in., beam 4½ in.). British and guaranteed—it's a HOBBIES-BOWMAN. The 'Snipe' exactly as in photograph costs only

22'6

(Postage 1/-)
Other models
from 17/6

For full details of the famous double-power 'Snipe' in the photograph and its sister ships 'Seahawk,' 'Eagle' and 'Swallow' (priced 42/–, 32/– and 17/6 respectively), write for our **FREE ILLUSTRATED FOLDER.** This folder also shows the clockwork launches and gives illustrations and specifications of the latest and one of the most successful Bowman inventions—the new, ultra-fast Aeroboat (see announcement on page 440 of this magazine). WRITE FOR *YOUR* COPY OF THIS INTERESTING BOAT FOLDER TO-DAY.

Bowman models can be inspected at all Halfords and Hobbies Branches and good shops everywhere.

BOWMAN models

BOWMAN MODELS (DEPT. MM3), DEREHAM, NORFOLK

An exciting and most successful advertisement for the *Snipe* steam launch from the *Meccano Magazine* of 1931.

strongly suspects, in Bowman workshops. All the stationary engines were horizontal ones—no vertical engines were ever made—and the simple single-acting oscillating cylinder was used exclusively. In the larger models the cylinders were cunningly disguised with fixed polished brass covers making the engines look not unlike gas engines.

A nicely preserved twin-cylinder Bowman stationary engine showing drip feed lubricators over the oscillating cylinders which are disguised by the polished brass covers.

The oscillating-cylinder engine, though not used in real practice for over a century, has in its simplest form much to recommend it as a toy power unit. It has no eccentrics or other valve gear, the oscillations from the main crankshaft providing movement between the cylinder block and the fixed steam chest. This enables the steam to enter and leave at the appropriate times in the cycle. In the diagram below (a) half-way down the working stroke the cylinder port and the steam port coincide fully. Full boiler steam pressure is then available on the top of the piston, pushing it down to bottom dead centre (b) at which point no further steam can get in. The momentum of the flywheel now carries the piston to the top of its stroke (c), exhausting the steam to atmosphere through the other port. Once over top dead centre the steam port is again in line with the cylinder port ready for another working stroke. The intermittent noise of the exhaust steam escaping gives the characteristic puff-puff, without which no steam-engine *is* a steam-engine. In all toy engines the steam is supplied at low pressure, often no more than 10–15 pounds per square inch, generated in boilers fitted with the basic feature of a filling/safety-valve which 'blows off' or releases the pressure should it rise too high.

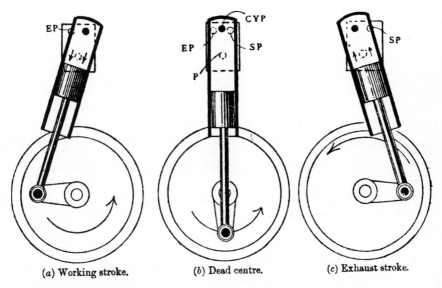

(a) Working stroke.   (b) Dead centre.   (c) Exhaust stroke.

The operating cycle of the oscillating engine used in very many steam toys.

The steam ports on the engines themselves are held in contact by springs strong enough to prevent steam leaking to atmosphere. However, in the unlikely event of the main boiler safety-valve sticking, the oscillating-cylinder engine itself provides a secondary safety-valve in that excess pressure will blow the port faces apart thus releasing steam and lowering the boiler pressure. This is one reason why these toy engines are so much safer than the slide- or piston-valve type still being made in Germany today.

19

The steam boilers are usually what are known as 'pot' type, that is to say, a simple closed vessel with a 'fire' underneath in a casing of some sort. The principle is the same whether it is designed to power a stationary engine, a launch or a locomotive. The commonest 'fire' is provided by a methylated-spirit lamp with cotton or asbestos wicks or, more efficiently, by a vapourizing lamp having a hot blue flame. Today there is a move towards the even safer solid fuels like Esbit or Meta, thus obviating bottles of easily flammable spirits. Some stationary engines made in Germany and the U.S.A. had (and have) electric immersion heaters in the boilers but to the purist this is really cheating and, anyway, most of the interesting smell is lost as well.

To return to the 1920s. At this time the popularity of Meccano construction kits was at its height. Meccano supplied clockwork and electric motors to drive the models but, although a steam-engine had been available for a very short time just before the Great War nothing comparable was sold by the company until 1929. Here then was an obvious marketing opportunity which Bowman Jenkins exploited. He drilled the metal bases of his engines with 5/32" diameter holes on ½" centres so that they could easily be incorporated into Meccano models. The advertisements made much of this and also the fact that they were drilled 'by kind permission of Meccano Ltd' though there must be a doubt about any proprietory right they had to a row of holes spaced ½" apart.

Big boilers, big lamps and well-fitting pistons and valve faces were important characteristics of Bowman engines. They worked; they worked well and long and were powerful for their size. The working instructions sent out with each engine were comprehensive and included a section on 'Rectifying simple troubles'. Should the engine stop suddenly, it noted, first of all look to see that the flywheel is not loose on the shaft. Many of us can remember many more likely causes than that!

In the 1927 *Bowman Book for Boys* Bowman Jenkins wrote:

In introducing our model steam engines for driving Meccano, Erector and other models we are opening a new era to the young embryo engineer. There is no pastime so interesting and beneficial to the would-be engineer as the making of models, but however successful his models may be, a tremendous amount of satisfaction and instruction is lost if he cannot drive them in exactly the same way as real machines. Many makers of really splendid models had never had the supreme joy of seeing them running under actual working conditions. In order to obtain the utmost realism steam power was obviously required but the foreign engines available were hopelessly underpowered and unreliable.

It was also realized that lubrication was vital to the efficiency of these small engines. First, it had to be the right oil and second, there had for preference to be drip-feed lubricators and felt oil-retaining pads. All this differentiated Bowman Jenkins' engines from the German ones he so despised.

Every model herein listed [he wrote], *even the very cheapest, is something very different from the accepted Continental production. Instead of fragile stampings sweated together, lead pistons and bright colour, nothing but the best British raw materials are used and British labour is employed throughout. All parts are accurately machined and screwed or bolted together. Pistons and cylinders of special metals are lapped to a dead fit and the valve faces are accurately ground to give perfectly steam tight joints. The result is that Bowman Models are real models and incomparably better in their marvellous performance and lasting qualities.*

It was all true too!

Lastly, remembering perhaps the occasional boiler explosion on the old test-tank, great care was taken in the design of the safety-valves which were by far the best of any fitted to toy steam-engines in their day.

The engines themselves ranged from tiny single-cylinder toys to quite sizeable twin-cylindered engines with geared countershafts and massive boilers. Such would run for over an hour and a half on one filling and the biggest would 'lift 112 lbs' or drive a sewing-machine. All had solid brass flywheels — no spoked wheels were used — and were mounted on flat steel bases or on wooden bases. By 1927 it was claimed that over 200,000 engines had been sold. Later a range of workshop toys was produced for them to drive, again with bases drilled for incorporation into Meccano workshops, where the framework and perhaps the overhead shafting would be used to provide the setting for the drills, saws and grinders. A generator, that indispensible steam-engine accessory, was also provided giving about 2 volts (enough to light one flashlamp) at speeds up to about 2,000 r.p.m. As one maker put it in Edwardian times, 'model engineers have long demanded electric light from the pent-up energy of methylated spirit through the medium of the model steam engine.' Plate 2 shows a Bowman engine, model number M135, coupled to one of the A.C. generators on a (modern) Meccano framework.

The addition of locomotives to the catalogue signalled a break with tradition, very characteristic of the man, once again designed to serve the purpose of sound engineering, powerful engines and safe working. By the 1920s the pre-war large rail gauges of 3 and 2 had ceased to be commercially viable and even gauge 1 was rapidly declining in popularity. Gauge 0 (1¼") was the favourite, but was soon to be challenged by gauge 00 which was too small for commercial live-steam locomotives.

The Bowman engines were made to run on gauge 0 track — tinplate or the more expensive forms of permanent way — but were built to gauge 1 size in every other respect. Here then was a sizeable pot-boiler, big (oscillating) cylinders and a lamp containing enough methylated spirit for at least a three-quarter-hour run. The express engine and tender (Plate 3) looked like a 4-4-0 but was in reality a 6-2-0 since no coupling rods were provided. This, and its sister 0-4-0 (or 2-2-0) tank engine was a particular favourite of Bowman Jenkins. He wrote:

An example of the unique qualities of the model 234 [the express] loco was demonstrated at the British Industries' Fair [probably 1925] when a standard model covered a total distance of 183 miles, the equivalent of a journey from London to Manchester, without special attention and without appreciable signs of wear. During the whole time it pulled six large coaches, the total length of the train being 10 feet.

It was re-fuelled every 40 minutes (i.e. a journey of 1½ miles). A £5 foreign engine refused to pull half the load and ran 12 minutes on one filling. Think of the joy an engine as just described can add to your own railway system.

And it cost, not £5 but £1 15s 0d (£1.75) complete with the tender.

The very size of these engines which made them something of an embarrassment on a small indoor track made them ideal for inexpensive garden railways. Soon a system of cheap weatherproof track (developed from another Bowman Jenkins patent, Tracks for Model Railways, No 331666 of 1928) was on the market. Rust-proofed rails simply fixed to rot-proofed timber battens and sleepers made possible a very satisfactory layout. To quote again:

Every mechanically minded boy desires his railway to be as much like the real thing as possible, and it is therefore only natural that the be-all and end-all of his designs and ambitions is an outdoor steam railway where real journeys can be accomplished, where he can dig out cuttings and tunnels, build embankments and bridges and generally cope in miniature with the problems of the railway engineer.

All good heady stuff despite the fact that there wasn't much real-life railway engineering of that nature left by 1930.

To match the locomotives, overscale coaches and wagons were made, the coaches with heavy bogies, wooden bases and ends with lithographed tinplate sides and roofs. All doors opened and the ingenious could easily remove the roofs and equip the interiors with appropriate seating, glazing etc. Despite the Bowman protestations about appearance the locomotives were poor representatives of any real-life prototypes. In fact one of his customers said they reminded him of Tin Lizzies (early Ford cars) in that they were mass produced by someone who, like Henry Ford, knew what was wanted and supplied it cheaply. But anyway, they worked like real ones, they sounded like real ones and, for those fortunate enough to possess one today will still run very satisfactorily—if you can find a really long 0 gauge track.

As we have seen the early 1930s were difficult years for luxury trades, and various world crises and the slump following the earlier Wall Street crash had made the business of toymaking a somewhat risky one. Bowman Jenkins tackled the problem from his point of view in two ways. From the toymaking aspect he diversified and the list of contents of the 1934 *Bowman Book for Boys* was in many ways remarkably similar to Richardson's of Liverpool catalogue of some sixty years before. There were, for instance, chemistry sets, water-motors and induction coils, together with aeroplanes (instead of balloons). He also introduced the Aeroboats; long thin speedboats equipped with slender cylinders containing, not compressed air, but elastic. This was a return to his earlier rubber-driven locomotive patent. Although they had some attractions, since the owner could decide to have long slow runs or short fast ones depending on how many strands of elastic he fitted, the Aeroboats didn't last long. The advertisements have a slight air of desperation about them at this time.

One of the first Hobbies engines made by Geoffrey Malins after Bowman Jenkins gave up. Something of the current Mamod appearance is to be seen in this early example though the geared countershaft is very much a Bowman feature.

Close-up of the Hobbies engine shown on the previous page.

Simultaneously he turned to the woodworking business in partnership with his cousin Captain Bernard A. Smart, who had some five years before sold his engines with Mr Warboys. Now, combining the names of Jenkins and Antique, he established the reproduction furniture company of Jentique Ltd, which is still flourishing in Dereham. The patent register is interesting for this period, reflecting as it does his wide-ranging ingenuity. It includes successful applications for patents for tables, stools, fittings for drop-leaf tables, means of winding up mechanical toys, razor-stropping apparatus and means for brushing teeth.

Perhaps due to his other interests Bowman Models was no longer functioning in 1935. Hobbies Ltd were not pleased at this withdrawal and started to look actively for someone to make the engines in his place. It so happened that another Geoffrey —Geoffrey Malins of Birmingham—had been supplying them with small metal components, notably propellers for their launches. He was approached and agreed to make some engines, closely following the Bowman designs, which Hobbies Ltd marketed under their own name. Thus the foundation was laid some forty years ago for today's range of Mamod toys.

Some time before the Second War Bowman Jenkins relinquished his association with Jentique Ltd and, living now in Ridlington in Norfolk, continued for a number of years in other businesses unrelated to steam-driven toys. About 1947 his ingenuity was directed to mass producing cheap pull-along toys made from a somewhat unlikely material—pyruma fire cement! At the same time a company using the Bowman trademark was making some steam toys in Luton, Bedfordshire. This we shall discuss in its proper place.

Geoffrey Bowman Jenkins died early in 1959 having been responsible during a comparatively short period of his life for giving an immense amount of pleasure to the children of his contemporaries and also some interesting and unusual engines for later collectors to cherish.

Chapter Three

MR G. H. MALINS
—THE EARLY YEARS

MR GEOFFREY HARRY MALINS, the founder and until his death in 1975, the Chairman of Malins (Engineers) Limited, was born near Birmingham in 1892 — just one year after Mr Bowman Jenkins. His father was the headmaster of a school in Bromsgrove in Worcestershire. Being of a practical turn of mind the youthful Geoffrey was apprenticed to an equally youthful organization, Mr (Sir) Herbert Austin's new motor car company. As a result of the training he received and the technical qualifications he gained he later became an Associate Member of the Institution of Mechanical Engineers. He was for a number of years the oldest living ex-apprentice of what is now the British Leyland Co's Longbridge works.

Some time before the Great War he decided that the sea had potentially more appeal for him and he entered the Merchant Navy. Here, after appropriate study, he was awarded a First Class Board of Trade Certificate as an Engineer and after the declaration of war in 1914 he transferred to the Royal Navy. He was for a time the chief engineer on a destroyer and saw some action. His ship had the misfortune to be torpedoed and as a result not long afterwards he was invalided out of the service.

Times were far from easy in the immediate post-war years. He returned to Birmingham, however, and here he married his wife Clarrie. She was a schoolteacher who had read English and graduated from Birmingham University with a B.A. and was to prove the greatest support to him in his early endeavours. She became, and indeed still is, the secretary of the company. But in the early 1920s it was her teaching and family affairs that largely pre-occupied her as their three children, all boys, came along; Eric the youngest being born in 1924.

Before this Geoffrey Malins put his engineering knowledge and experience with large marine steam plants to use in a land-based job. This was a time when both the domestic and industrial use of electricity was developing rapidly. Extensions were being made to municipal supply undertakings and a number of the larger industrial concerns were establishing their own private generating stations. One of these was the Dunlop Rubber Company at its Fort Dunlop works in the north of Birmingham. Here Geoffrey Malins was for some time in charge. He soon established a reputation for fanatical efficiency and made what might well be considered today to be unreasonable demands on both the men and the machines under his care. The station in fact sold something like half its output of electricity to the Birmingham Corporation for the public service—a situation not uncommon among the private generating stations at that time.

His first venture into self-employment was a garage business on the Tyburn Road, Birmingham, which lasted for a year or so. It was not wholly successful and he then embarked upon the venture which was ultimately to lead to the steam-toy business. In 1934 he set up on his own in a single room 'factory' at No 26 New Buildings, Price Street in the middle of Birmingham. This was in what was known as the 'Gun Quarter' where traditionally the workers in the gun trade had for centuries operated a highly skilled and highly individual system of small workshops, each one specializing in one component of the finished weapon. Geoffrey Malins was not therefore unusual in having his own one-man shop in the area. He did not, however, make gun components. As far as we can discover his first product was a small ball-race turntable which he proudly claimed could carry 56 lbs in weight. It was first advertised in the *Meccano Magazine* for November 1934 as being most suitable for building into Meccano models. Later a larger version was made to carry 1 cwt. The 'firm' was known as the 'G. M. Patents Co' though no patents have been found in Geoffrey Malins' name. The small turntable sold for 1s 6d (7½p) and, trade

"G.M."
Ball Race Turntable
∴ Competition ∴

1st. Prize £2-2-0
2nd. Prize £1-1-0
: : and 12 : :
Consolation Prizes

The above prizes are offered for the best model or apparatus in which a "G.M." BALL RACE TURNTABLE is incorporated.

Send in a photograph or accurate drawing of your model or apparatus to:—

"G.M." PATENTS CO.,
26 New Buildings
Price Street, Birmingham
and enclose the PINK WRAPPER (without which, entry is disqualified)

In all matters relating to this Competition, our decision must be accepted as final.

Left and overleaf:
Geoffrey Malins' first recorded product as advertised in the *Meccano Magazine* in 1934.

The "G.M." Ball Race
TURNTABLE

PATENT APPLIED FOR

For Universal Adaptation to Turn Weights Up To 56 lbs.

Suitable for any toys such as — MECCANO and TRIX which appeal to the Modern Boy interested in mechanical construction.

The "G.M." Ball Bearing Turntable is a thoroughly reliable engineering production, accurately made and contains 45 Hardened Steel Balls with allowance for movement — a self contained action and should not be dismantled.

Stove Enamelled in Post Office Red

If necessary, the Ball Race may be adjusted by the centre screw after unlocking the nut

SECTION VIEW
Illustrates the great strength of the turntable and shows the method of adjustment

The "G.M." TURNTABLE has many other uses which will readily be appreciated both by Manufacturers and practical Amateurs

PRICE 1/6

apparently proving less than brisk, some of the traditional expedients were tried. First there was an appeal to the retailers to order early 'before stocks are depleted' and later a competition was used to drum up more business.

Other products traced to this period were some rather superior brass propellers for model boats. The British Industries Fair, then held annually on a permanent site at Castle Bromwich, not far from Fort Dunlop, was a favourite exhibition centre for firms large and small. We have seen in Chapter II that Bowman Jenkins first showed his steam locomotive at a Fair in the 1920s. Now, a few years later, Geoffrey Malins also had a stand at the Fair and there he met the directors of Hobbies Ltd who were fellow exhibitors. His propellers and possibly some other components were of interest to them and soon 'G. M.' products began to appear in the annual Hobbies handbook. It was not long after this that, Bowman Jenkins having given up toy steam-engines, Geoffrey Malins was asked to make a small range which Hobbies Ltd sold under their own name.

In the 1938 (published October 1937) edition of the *Hobbies Handbook* these early 'Malins/Hobbies' engines are listed and illustrated together with the remaining Bowman stock of *Swallow* and *Peggy* steam launches and a 'set' for boys to build an engine themselves which was probably made up of old Bowman components that had been left over at Dereham!

The new engines resembled the old in many ways to maintain the tradition. But whereas Bowman (doubtless under Hobbies' influence) had offered each stationary engine with a wooden baseplate as an alternative, the Malins engines were only supplied with flat, metal bases drilled for Meccano. Again, the old engines had various brass chimneys set on spigots on the baseplates. The new had what were later termed 'locomotive type' chimneys fixed to the boilers, a feature which is continued to this day. We can see this as an early example of value analysis since the same purpose was served by the new chimney at a saving of over half the length of the brass tubing.

The steam launches were abandoned and after the Bowman stocks were exhausted no more were made or sold by Hobbies Ltd. Some of the launch engines, notably that used in the *Peggy*, had been sold separately for fitting into home-designed and made boats and it was not long before Geoffrey Malins introduced a similar simple marine steam plant complete with stern tube and propeller. This was the M.E.1.

Another casualty was the gauge 0 locomotive range. Although the Bowman 4-4-0 was a considerable technical success it did not look very much like a real locomotive with its obtrusive oscillating cylinders and lack of any connecting rods. Furthermore in 1931 Bassett-Lowke Ltd had introduced their first post-War 4-4-0 locomotive *Enterprise,* and a very efficient and handsome looking engine it was (Plate 4). Although it sold for £2 10s 0d (£2.50) complete with tender as against £1 15s 0d (£1.75) for the Bowman it was clearly a much more attractive toy. It is of some interest to compare the performances quoted in the rival advertisements and testimonials for these two engines. The Bowman's six coaches were countered by Bassett-Lowke's claim for their engine to pull five long bogie coaches. A Bowman customer in London recorded a running time of 50 minutes non-stop (about a mile and a half) and Bassett-Lowke replied with a claim for a run of 55 minutes covering

1¾ miles. Certainly the *Enterprise* in the photograph still runs splendidly after many years and it will tick over very very slowly most beautifully. One customer got one to run light for 1 hour 10 minutes non-stop. Malins did not however produce locomotives and indeed have not made any to this day.

Until the 1939 war stopped toy production Geoffrey Malins remained in his small gun-quarter workshop, doing things the hard way as his son Eric puts it, now with two or three part-time women to help. His production capacity was slightly increased during this time by the use of part of the family home, an old Citroen car engine being rigged up as a power source to run some of his lighter machinery. Compared to the advanced production methods used today in the Brierley Hill works it is almost unbelievable that so much was achieved by hand and simple machine methods in small batch production quantities. But this was true of many small Birmingham manufacturers and especially true of the toy trade, which had for many years an unenviable 'back shed' image.

In those days the engine frames were made from brass castings which were brought out in a semi-finished state. The flat steel baseplates were clamped together in small quantities and drilled on a simple drill press for the engine and boiler fixing holes *and* the rows of Meccano holes as well. The turned parts — cylinders, pistons, flywheels etc were machined on a Ward la capstan lathe that survived two moves to the present works and was kept (perhaps for sentimental reasons) until it was scrapped quite recently.

Although Hobbies Ltd had been the sole outlet for the Malins engines for the first year during 1936/37 the next step was the establishment of the company's own name 'MAMOD' and an independent marketing system which nevertheless still included Hobbies. The first four-page leaflet (undated but thought to be 1937) to bear the name 'MAMOD' does not carry any address of the company. It does, however, list all the engines made before the 1939–1945 war and includes for the first time details of two sizes of line-shafting and three working models for the engines to drive. The models are simple; a polishing spindle, a power hammer and a power press, all on Meccano drilled bases and all 'strongly made of castings (iron)' and, in the case of the polishing spindle, 'with real felt polishing bobs'.

The engines illustrated, though strongly reminiscent of the Bowman range, are in general smaller and more compact. There is the beginning of rationalization — of baseplates for example. With the exception of the Minor No 1 there are two sizes of plates for four engines. Four models (including the marine engine) have heavy brass crank discs which also act as the flywheels; only the two largest engines have separate solid brass flywheels. These are also the only two fitted with steam whistles and packed in strong plywood boxes, which was another Bowman/Hobbies feature. The cheaper ones are packed in cardboard! All are fired by methylated spirits burned in simple lamps fitted with asbestos wicks; two, three or four burners according to boiler size.

Opposite and following three pages: The first four page catalogue issued by Mamod about 1937. It carries neither a date nor an address but the company was at this time in Price Street, Birmingham.

Model Steam Engines
and Working Models

MADE BY ENGINEERS

MODEL S.E.4   STATIONARY STEAM ENGINE

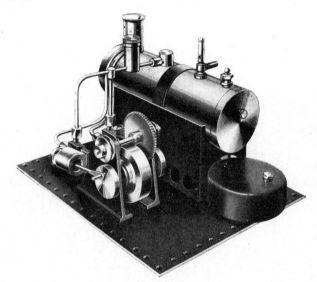

This engine is fitted with two cylinders and geared reduction shaft. It is built on heavy cast brass frames, with long lubricated bearings, a heavy solid brass flywheel, turned brass crankdiscs and pulleys. A driving pulley is incorporated in the flywheel.

The cylinders are lubricated by a lubricator in the steampipe.

The boiler is of solid drawn brass tube, with domed ends, 6″ long, by 2″ diameter, fired by a five wick spirit lamp with asbestos wicks, and is guaranteed to 100 lbs. pressure. It is fitted with brass dome and union, brass spring loaded metal seat safetyvalve,

whistle and waterlevel plug. A brass loco type funnel is fitted on the boiler, to which the exhausts of the engine are led for realism as in large practice.

This engine is mounted on a steel baseplate 7½″ by 6½″, drilled at the edges to fit Meccano, and enamelled with oil-resisting signal red enamel.

This is a very powerful engine, which runs at 2,000 revolutions per minute.

The C.2 countershaft fits this engine, as also S.E.3.

Packed in strong plywood box, complete with filler funnel and instructions.

PRICE, 29/6   POSTAGE, 8d.

MAMOD MINOR No. 1

A well made engine, which fits Meccano baseplate $4\frac{1}{2}''$ by $2\frac{1}{2}''$, strong brass boiler fitted with safety valve, heavily made brass engine frame and cylinder. There is no lead used in this engine.

A low priced addition to the famous range of Mamod engines, acclaimed the best on all markets. Height of engine $4\frac{5}{8}''$, packed in strong cardboard box, complete with filler funnel and instructions.

PRICE, 5/- POSTAGE, 6d.

MODEL S.E. 1 STATIONARY STEAM ENGINE

A very strongly made steam engine on best engineering lines.

Boiler of solid drawn brass tube with domed ends for great strength. It is $1\frac{3}{4}''$ diameter, by $3\frac{1}{4}''$ long, and guaranteed to a pressure of 100 lbs per square inch. Is fitted with solid brass safetyvalve with metal seat.

The engine has a cast brass frame, long bearing with lubricating hole. The cylinder is brass, bored from the solid, the flywheel and pulley of turned brass.

The complete engine is mounted on a steel baseplate $5\frac{1}{2}''$ square, drilled all round to fit Meccano.

The engine is finished in attractive colours—signal red baseplate, engine frame green and brass, boiler polished and lacquered, funnel polished brass, boiler cradle black.

Packed in strong cardboard box, complete with filling funnel and instructions.

The C.1 lineshift fits this engine.

PRICE, 10/- POSTAGE, 6d.

MODEL S.E. 2 STATIONARY STEAM ENGINE

This well-designed engine has solid drawn brass boiler with domed ends, 4" long by $1\frac{3}{4}''$ diameter, a brass spring loaded safetyvalve, and a waterlevel plug is fitted in the end of the boiler. A loco type brass funnel is fitted, connecting with exhaust from the engine for realism.

The engine is fitted with reducing gear and fits Meccano pulleys.

The engine frame is of cast brass, long bearings with oilholes are fitted, and the cylinder is bored from the solid. Steam and exhaust pipes are connected up with hexagon unions (brass).

The complete engine is mounted on $5\frac{1}{2}''$ square steel baseplate, drilled at edges to fit Meccano, which is enamelled signal red, oil resisting, the engine frame is enamelled green, the boiler cradle black, and the boiler, funnel, cylinder, polished brass and lacquered.

Packed in strong cardboard box, complete with filler funnel and instructions.

The C.1 lineshift fits this engine.

PRICE, 14/6 POSTAGE, 6d.

MODEL S.E.3 STATIONARY STEAM ENGINE

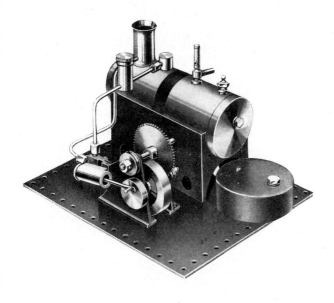

The engine is built on solid brass frames, long bearings, with lubricating holes. The cylinder is bored from the solid brass, and flanged as in large practice. The reduced gear shaft fits Meccano, is driven through machine cut gears. The flywheel, crankdiscs, and pulleys are of turned brass.

The engine exhausts to funnel as in large engines.

The boiler is of solid drawn brass tube, 5″ long, by 2″ diameter, with domed ends, and is guaranteed to 100 lbs. pressure. It is fitted with brass dome and union, brass metal seat spring loaded safetyvalve, whistle, and endplug to determine the correct waterlevel when filling. A loco type funnel (brass) is fitted.

A lubricator is fitted in the steampipe for the best lubrication of the cylinder. All steam and exhaust pipes are of best copper, connected with hexagon brass unions.

The four wick spirit lamp is fitted with asbestos wicks to prevent charring, and ensure best performance of the engine.

A steel baseplate, 7½″ by 6½″, drilled at the edges to fit Meccano, and enamelled signal red, oil resisting, mounts the complete engine. The engine unit is finished brass and green, the boiler, funnel, and cylinder are polished brass and lacquered. An engine of first class engineering design and construction, and of very pleasing appearance. Packed in plywood box, complete with filler funnel and instructions.

C.2 lineshaft fits this engine.

PRICE, 21/- POSTAGE, 7d.

MODEL M.E.I MARINE STEAM ENGINE

A well-made model marine engine. Has a cast brass engine frame, with cylinder bored from the solid, heavy turned brass flywheel. The propellor shaft is brass, with polished cast brass propellor, pitched as in large speedboat practice. The complete engine unit with stern tube can be removed from the baseplate and mounted separately in your boat.

Boiler of solid drawn brass tube with domed ends, fitted with spring loaded brass safetyvalve, steam dome and union.

Mounted on steel baseplate 12″ by 3″.

PRICE, 12/- POSTAGE, 6d.

The manufacturers retain the right to alter specifications and prices of all goods without notice.

LINESHAFT C.1

Strongly made with long bearing lubricated pedestals of cast iron, and silver steel shaft $5\frac{1}{2}''$ long. The height to top of pedestal is $3\frac{1}{2}''$.
Four pulleys are fitted, $\frac{1}{2}''$ and $\frac{3}{4}''$ brass pulleys, $1\frac{1}{4}''$, $1\frac{3}{4}''$ pulley. For convenience in fixing is mounted on a steel baseplate, 6″ by 4″.

Fits, when taken from steel baseplate, the S.E.1 or S.E.2 engines, and fits Meccano. The shaft fits Meccano pulleys.
Attractively finished and packed in a strong cardboard box.

PRICE, 3/6. Postage, 6d.

Extra pulleys and pedestal bearings can be supplied, see Spares list.

LINESHAFT C.2

Larger than C.1, and fitted with six pulleys, $\frac{1}{2}''$ and two $\frac{3}{4}''$ brass pulleys, two $1\frac{1}{4}''$ and $1\frac{3}{4}''$ built up pulleys. This lineshaft fits Meccano pulleys, and is $6\frac{1}{2}''$ long.
The pedestals are of cast iron with long lubricated bearings, are $3\frac{7}{8}''$ high, and are mounted on a steel baseplate, 7″ by 4″, with edges drilled for Meccano. This lineshaft fits the baseplates S.E.3 and S.E.4 engines, or Meccano. Attractively finished, is a model of large engineering, and packed in a strong cardboard box.

PRICE, 4/6. Postage, 6d.

Extra pulleys and pedestal bearings can be supplied to enable special layouts to be made up, see Spares list.

M.1. POWER HAMMER.

An accurately designed model, heavily constructed in cast iron. Fitted to baseplate, which is drilled to fit Meccano.
Our S.E.1 engine will drive three of these hammers.
Baseplate $2\frac{1}{2}''$ square, total height full 2″.
Attractively finished in red, green, and black.

Splendid Value, 1/9. Postage, 3d.

Packed in strong cardboard box.

M.2. POLISHING SPINDLE.

Total Height, $2\frac{3}{4}''$.

Built of castings (iron), with real felt polishing bobs, $1\frac{1}{4}''$ diameter. The baseplate is $2\frac{1}{2}''$ square, drilled all round to fit Meccano. Painted green, with signal red baseplate. Packed in a strong box.

PRICE, 2/6. Postage, 3d.

M.3. MODEL POWER PRESS.

Strongly built of castings (iron). With long lubricated bearings. This can be easily driven by our smallest engines. Press body painted green, and mounted on signal red baseplate, drilled to fit Meccano.
Total Height, 4″.
Packed in a strong box.

PRICE, 2/3. Postage, 3d.

SPARES LIST

BOILERS.

		s.	d.
S.E.1.	Boiler, complete with safety valve and union nut ...	4	10
S.E.2.	As above	6	3
S.E.3.	As above	8	6
S.E.4.	As above	9	6

ENGINE FRAMES.

		s.	d.
S.E.1.	Complete with steam pipe	2	0
S.E.2.	Complete with union nuts	3	9
S.E.3 & S.E.4 ...	Prices on Application		

CRANK SHAFTS
(complete with flywheels).

		s.	d.
S.E.1 and S.E.2	... ...		8
S.E.3	... ...	1	9
S.E.4.	Incorporating $\frac{3}{4}''$ pulley ...	2	3

PULLEYS AND GEARWHEELS.

	s.	d.
$\frac{1}{2}''$ bored $\frac{1}{8}''$		4
$\frac{1}{2}''$ bored (Meccano size)		4
$\frac{3}{4}''$ bored (Meccano size)		5
$\frac{3}{4}''$ bored $\frac{1}{8}''$		5
$1\frac{1}{2}''$ gearwheels (Meccano size) ...		7
$\frac{1}{2}''$ pinions, bored $\frac{1}{8}''$		4
$1\frac{1}{4}''$ pressed, bored (Meccano size)		4
$1\frac{3}{4}''$ pressed, bored (Meccano size)		5

SAFETY VALVES.

	s.	d.
Complete, to fit all engines ...	1	3
Springs for valves		2
Washers for valves ... per dozen		6
Sockets for valves		2

UNIONS.

	s.	d.
Whistles, complete	1	9
Unions, complete with washer ...		5
Union nut		2
Union screwed cone		2
Union washer		1
Lubricator Cups		2

SPIRIT LAMPS.

	s.	d.
2 burner, complete with wick ...	1	6
4 burner, complete with wick ...	2	6
5 burner, complete with wick ...	3	0
Vented plugs for all sizes		2
Sockets for plugs		$1\frac{1}{2}$
Wick asbestos		4

END PLUGS.

	s.	d.
End plugs		2
Socket		$1\frac{1}{2}$
Washers ... ... per dozen		6

CYLINDERS.

	s.	d.
For S.E.1 and S.E.2, complete with trunnion spring and nut ...	1	9
For S.E.3 and S.E.4, complete with trunnion spring and nut ...	2	3
S.E.1 and S.E.2 spring and nut ...		$2\frac{1}{2}$
S.E.3 and S.E.4 spring and nut ...		$2\frac{1}{2}$

	s.	d.
FILLER FUNNELS		2

CRADLES.

	s.	d.
S.E.1 and S.E.2		9
S.E.3	1	0
S.E.4	1	3

CRANK DISCS.

	s.	d.
For S.E.3 and S.E.4, complete with pin and screws		8

PEDESTALS.

	s.	d.
For C.1	1	0
For C.2	1	3

There is still a heavy reliance on brass castings for the engine frames and the geared countershafts whilst the cylinders themselves are made from solid bar stock. The leaflet proudly states that the engines are British made and also that they are 'Made by Engineers' a statement borne out by extensive use of screwed unions for the steam and exhaust pipes and efficiently arranged lubricators for the larger engines.

Whilst the Mamod engines were probably the most popular and sought after of the British steam toys sold just before Hitler's war they were not, of course, the only ones. Among the competitors were the high quality Bassett-Lowke stationary engines and a wide range of engines available from Stuart Turner Ltd of Henley-on-Thames. These were mostly supplied as sets of drawings, materials and castings for model engineers to build in home workshops and thus not strictly within our meaning of Toyshop Steam. However, some of the smaller and cheaper oscillating-cylinder engines and boilers were really toys and comparable with the Mamod products.

This is a reproduction of the famous *Henley* launch made by the author in 1975. The original was produced by Stuart Turner Ltd in the 1920s and 1930s. The steam plant is still being made today over fifty years since it was first designed.

As we have seen, many of the makers were in a very small way of business indeed, and few put their names to their goods. Among those who did was the Mersey Model Company of Liverpool. They made a range of four oscillating-cylinder stationary engines of a rather simpler design than the Mamod toys, though the largest was fitted with a belt-driven countershaft. The company claimed to have been making engines for over thirty years, but it has not been possible to trace them in trade directories before 1937 (when they were at New Brighton) nor after 1941. The photograph (Plate 11) also shows two of their attractive die-cast machine tools, an eccentric press and a lathe, which they made for the engines to drive.

The Clyde Model Dockyard of Glasgow, which was shortly to cease trading after about a hundred and forty years, brought out the curious steam-driven outboard

"S.T. TWIN"

New design incorporating twin S.T. Cylinders. A splendid engine for boats or driving models, etc. Fitted with Steam Cock for start, stop, and reverse.
Double-acting cylinders at 90°—no dead centres—practically self-starting. Set of Parts, all machined, drilled and tapped, with instructions,

13/6

Finished   Engine, painted and **17/6** tested,

Hamleys
Estd 1760
·HAMLEY·BROS·LTD.·

Branches : London and Croydon.

Mail Order Depart- ment (M):
200/202, REGENT STREET, LONDON, W.1.

Left and below: Advertisements from the 1930s of two of Mamod's competitors.

Make your own Speedboat.
Use this
MARINE ENGINE

Well made and strongly constructed Marine Plant suitable for Boats approximately 24" in length. Fitted with safety valve and exhaust pipe over boiler giving the appearance of smoke from chimney. 2-Wick Burner tested under steam prior to leaving works. Length 5", width 1¾", depth 3½". We have a large selection of hulls. Post 6d. Price **7/3**

marine engine shown on page 39. This was in 1932. It came as a complete unit for fitting to home-made boats and had a neat brass case with bevel drive to the two-bladed propeller. A boiler unit, possibly the one shown in the picture, was also made by the company and was fitted with a light aluminium housing and a rather crude methylated spirit lamp.

Bond's o' Euston Road, London, sold a few British-made engines particularly those designed for use in launches and boats and W. H. Hull and Son of Birmingham (established in 1884 and a Mecca for generations of Midland schoolboys) advertised oscillating-cylinder tank locomotives in gauge 0 as late as 1927. Gamage's annual catalogue, issued to the great joy of many of us when young just before Christmas, also proudly announced that some of their engines were made in Britain. Again, a marine plant is remembered. Since the majority carried no identification marks at all it is now extremely difficult to be sure of the origin of

Perhaps one of Hull's Gauge 0 engines (much re-built) is here in steam heading a goods train of Hornby rolling stock.

COLOUR PLATES

Plate 1   *Fury* is a tinplate 'Birmingham Dribbler' locomotive inspired by the Great Western Railway's broad-gauge engines and made perhaps in the 1880s.

Plate 2   Even the single-cylinder Bowman engines were capable of driving a small electric generator.

Plate 3   Bowman's much-loved locomotive. Nominally 4-4-0 it was really a 6-2-0 since no coupling rods were fitted. The coach has a wooden base and ends with nicely lithographed tinplate sides and opening doors. Although running on gauge 0 track all Bowman trains were otherwise to gauge 1 standards.

Plate 4   Bassett-Lowke's elegant rival to the Bowman locomotive. This is *Enterprise*, a 4-4-0 live steam-engine in gauge 0, first introduced in 1931. Here it is hauling a train of Hornby Pullman cars through Beaconsfield station—a tinplate edifice made by Georges Carette in Nuremberg about 1915 and also sold here by Bassett-Lowke.

Plate 5   The majority of simple engines had no makers' nameplates. This vertical engine, with a single lapped brass boiler and crankpin set in the brass flywheel looks English though it carries no identification. It is driving a simple water toy and a smith in an unlikely half-timbered workshop, both of German origin from the mid 1920s.

Plate 6   A fleet of P.W.E. (Pulsating Water Engine) boats. The largest was made by Jetcraft Ltd. near Manchester and the blue speedboat with driver is an original German Toc-Toc boat of about 1928.

Plate 7   A nicely preserved example of Mamod's *Meteor* launch—the only model that the company possesses.

Plate 8   The Mamod Steam Roadster.

Plate 9   The Mamod SR 1(a) Steam Roller and the TE 1(a) Traction Engine with their trailers.

Plate 10   The full range of Mamod stationary engines.

Plate 11   In addition to a range of stationary engines the Mersey Model Company made attractive die-cast machine tools for them to drive. This photograph shows an eccentric press and a lathe.

Plate 12   Mamod's twin cylinder SE 3 superheated engine driving a toy workshop comprising all the company's accessories.

Plate 13   The Mamod Steam Wagon in its new livery.

3

4

8

9

11

12

In 1932, towards the end of its long life, the Clyde Model Dockyard marketed this curious steam outboard-engine with an aluminium cased boiler. It was sold for fitting into customers' own hulls. Here it is installed in an earlier mahogany boat that is very much too heavy for it.

Close-up of the above.

BRITISH MADE

STEAM LOCOMOTIVES

0 Gauge Loco. Brass boiler, brass oscillating cylinders, anti-friction pistons, automatic oiler, wind guards. Will steam about 15 minutes with one filling. L.M.S. or L.N.E.R. colours.

Price 12/6

Postage **6d.**

Strong heavy loco, 0 Gauge, brass boiler, brass oscillating cylinders, anti-friction pistons, automatic oiler, wind guards. Will run for half mile with one filling. Exhaust through the funnel.

Price 21/-

Postage **6d.**

These locos have given every satisfaction.

W. H. HULL & SON

MODEL RAILWAY SPECIALISTS

North Western Arcade, Birmingham

Established 1884.

Hull & Son, established in 1884, was for many years a popular store in Birmingham for generations of Midland schoolboys.

many of those that survive. Plate 5 shows an example of a very simple vertical engine of the period with a single soft-soldered lap joint on the boiler and a sand cast aluminium base.

The threat of war in 1938 and the ensuing uncertainties for business were only resolved with the declaration of war against Germany in the late summer of 1939 — for the second time within twenty-five years. It was in this year that Geoffrey Malins demonstrated his faith in his products by again taking a stand at the British Industries Fair to show the full range of his engines. Demand increased dramatically almost at once but by Christmas the Mamod toy business was put into cold storage 'for the duration'.

The Price Street workshop was retained, however, and Geoffrey Malins undertook some minor production work to help the war effort. But this was not enough to

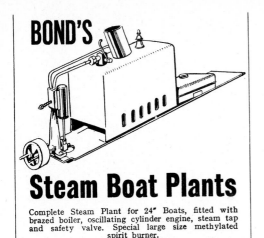

BOND'S

Steam Boat Plants

Complete Steam Plant for 24″ Boats, fitted with brazed boiler, oscillating cylinder engine, steam tap and safety valve. Special large size methylated spirit burner.

Price 10'6 POSTAGE 9d.

If you want anything for your **Model Railway, Model Boat** or **Aeroplane**, send for Bond's Catalogue, **6d.** post free. This also includes a full list of Tools.

Bond's O' Euston Road Ltd.
254, Euston Rd., London, N.W.1.
'Phone: Museum 7137.   Estd. 1887.

Another advertisement from the 1930s of one of Mamod's competitors.

satisfy him and, now in his late forties, he again found himself at sea—literally so—as an engineer helping to deliver small vessels, such as minesweepers, to destinations as far afield as Malta. His eldest son, Philip, joined the Army and the next, Clive, who had already started to take an active part in the business, joined the Fleet Air Arm. He was to be tragically killed in a later phase of the war. His youngest son Eric followed his father into the Navy. When peace came again and the toy business was revived it was Eric who took over the role in the company intended for his elder brother.

c. 1930

Chapter Four

REVIVAL AFTER 1945

WHEN THE END of the war was announced in 1945 on May 8th (VE Day—for Victory in Europe) life slowly started to return to something approaching the normality of the 1930s. Another three months were to elapse before VJ Day (Victory in Japan) on August 17th, after which we were no longer at war with anyone. The effects of the struggle lasted even longer than in the 1920s. There was a grave shortage of goods, services and food. Petrol was severely rationed until 1949 and food rationing was not finally abandoned until nearly ten years after the war. Raw materials of all kinds for industry were difficult to get; electric power cuts were frequent, and steel and non-ferrous metals in particular were in very short supply.

Nevertheless everyone tried hard to re-establish businesses put into cold storage due to the war, or return to their traditional products after helping to make Spitfires, tanks, guns or shells. The metaphorical beating of swords into plough-shares was done with great energy and this resurgence was very marked in the Midlands, not least among small family firms such as Malins.

Malins had a number of advantages. First of all Geoffrey was joined by his youngest son Eric who had been trained as a cost accountant. Then some of their female part-time staff came back to the firm (including one who is still with them). Then again they had retained their premises and acquired part of an old gun factory almost next door in St Mary's Row in the centre of Birmingham. This was in a Georgian house behind the General Hospital and the top floor was used both as an office and a workshop. All is now demolished. Eric Malins remembers with amusement that it was flanked on one side by a Convent and on the other by a working girls' hostel. Other neighbours more appropriate to the business included toolmakers, stampers and brassfounders. And lastly they had retained all their design drawings and especially the tooling which had been used to produce the pre-war engines.

Eric took on the managing directorship with his father as Chairman and the first engines they made in 1946 were the Minor I—the tiny overtype model—the SE I, followed almost immediately by the SE 2. These were virtually the same as the pre-war versions, the SE 2 having a geared countershaft but now also fitted with a throttle valve on the exhaust pipe to control the speed. For a very short time, about 1935 or 1936, there had been a twin opposed-cylinder overtype engine called the

Length 24½ in., Beam 5 in.   All metal and alloy construction.

Moulded on the lines of the very latest post-war motor torpedo boats, the magnificent design, high efficiency power unit, ultra modern appearance and superb finish make **METEOR THE FINEST VALUE EVER OFFERED IN MODEL STEAM YACHTS**

£3 - 18 - 4
retail
inc. Tax

Be sure to see this and and other fine productions on

STAND N 122, B.I.F., OLYMPIA

Manufactured and Fully Guaranteed by

MALINS (ENGINEERS) LTD
2 ST. MARY'S ROW, BIRMINGHAM 4

Telephone:
CENtral 6175/6

Telegrams:
Mamod, Birmingham

An early *Meteor* advertisement in a 1949 issue of the trade journal *Games and Toys*.

Modern Mamod counter display.

Minor 2, but this disappeared before the war and the new version, substantially the same as the current model, was not introduced until about 1949. The pre-war SE 4 was not made again and the present SE 3 is a totally different engine to the one in the 1937 catalogue. Some of the engine mountings and gear frames were still being made of brass sand castings and, due to the prevailing shortages, the small deliveries which the suppliers were able to make had to be saved up until enough were in stock to justify setting up to machine a batch. And in the disasterously cold winter of 1946/7 with fuel short, power limited and coal piling up in all the wrong places even the machining was unpredictable.

A move was made to up-date their methods and some hot-brass stampings were used in place of sand castings. Similarly the accessories or working toys—the hammer, press and polishing spindle—were still made of cast iron for a time until a change to Mazak (a zinc based alloy) was made in 1950. Few companies managed to achieve anything like their full output in these immediate post-war years, and there was therefore pressure on all managements to overcome some of the limitations by design changes and other developments to improve efficiency. About this time the geared countershafts were abandoned and for the first time spoked flywheels were adopted. These were made from hot-brass stampings, initially from a tool acquired from the toolmakers who had supplied flywheels from it to the Mersey Model Company before the war.

The St Mary's Row premises were not very satisfactory—it was necessary for instance to rely on a hand-operated 10 cwt goods lift—and in 1948 a move was made to new works in Camden Street, Birmingham, not a thousand miles from James Watt's Soho Foundry. Here there was a floor area of some 3,400 square feet and by now about forty people were employed. Though toy engines were the sole product, the demand tended to be seasonal, with a heavy concentration on the Christmas trade. So a new product was adopted as a bread-and-butter line. This was a range of ball-catches for the hardware trade. It was the only non-toy product Malins ever made and after a dozen years or so, in 1962, it was abandoned and the tools sold.

The year after the move, in 1949, the first attempt to break away from stationary steam-engines was made in the shape of the *Meteor* steam yacht, perhaps also with the idea of capturing some summer trade and evening out the annual work load. The *Meteor* was an attractive vessel about 25" long × 5" beam with a substantial pressed steel hull. The hull was bought out from a firm in Worcester but the engine, an early version of the post-war ME 1, together with the fittings were made by the company. The deck housings were well designed for ventilation of the spirit lamp under the boiler and the two sliding hatches gave rather restricted access to the engine. She was painted in brown, cream and red and came, as the instructions stated, 'packed with spirit lamp, flag and mast'. The young pond-side mariner was, however, encouraged to take with him a store of additional equipment to add to his 'Convenience and Pleasure':

i) A tin or bottle for water. NEVER fill the boiler with pond, soapy or dirty water. The cleanest tap water only must be used.

ii) *A tin or bottle for methylated spirits (known in Canada and U.S.A. as medical alcohol). NEVER use paraffin.*

iii) *An oil can containing medium oil for filling the engine lubricator.*

iv) *An old walking stick for controlling the boat in the water.*

v) *A reel of thin fishing twine or thread for attachment to the boat if sailing in deep water.*

vi) *A duster or cloth for wiping the spirit lamp and drying the hull after use.*

The *Meteor* sold for £3 18s 4d (£3.92) later increased to £4 4s 0d (£4.20) but was not a successful venture. She worked well but possibly was too expensive for the market in those days. On the other hand, the engine was completely hidden when sailing and, since you couldn't see the wheels go round, much of the visual appeal of a live-steam boat, evident in the earlier Bowman launches, was lost. It is ironic that one of the company's very few commercial failures was the only steamboat designed and made by two ex-Navy men! Less than 1,500 were made and in consequence they are comparatively rare today.

The same hull was also fitted with an electric motor, in place of the steam-engine, particularly aimed at the American market. In this form she was named *Conqueror* but, selling at the same price she fared little better commercially.

As explained in Chapter Two practically all the British toymakers used the oscillating-cylinder engines for cheapness, simplicity and their use as secondary safety-valves. There is, however, one form of propulsion adopted for small toy boats that does not use a conventional engine. It is nevertheless a true heat engine and should properly be mentioned here in connection with toyshop steamboats. Boats fitted with these simple devices are variously known as flash-steamers, hot-air-boats, toc-tocs (after a German version in the 1920s) or pop-pops. A more accurate term for the 'motors' is Pulsating Water Engine (P.W.E.) since this is what they are. The pulsations are induced by a small spirit lamp and there is a typical collection of the boats shown in Plate 6.

In its simplest form the 'engine' consists of a short length of metal tube bent into a U, with the two open ends projecting from the stern of the boat just below the water line. The top of the U (or sometimes just a simple coil) is amidships and it is under this that the lamp is placed. To start, the tube is filled with water, the lamp lit and after a few seconds it heats the water in the bend or coil of the tube enough to generate a little steam. This expels the water in the tube with sufficient force to give the boat a slight push forward. Immediately the cold water outside condenses the steam and it rushes back to the hot part of the tube to be again turned into steam, so repeating the cycle. This takes place quite rapidly, the pulses making a phut-phut noise as the water bubbles out, driving the boat steadily forward by reaction.

Such simple 'engines' are easy to make and despite statements to the contrary there is no mystique about the form of the U or the coil. All that is critical here is the size of the flame. Too much or too little heat and the pulsations will fail to start or to continue. Also the tube ends must lie just below the water line.

The earliest P.W.E.s that are known to the author originated in Germany before

1914 in a series of boats made by one of Nuremberg's most imaginative and prolific toymakers, Ernst Plank. Perhaps the first in England, and certainly the best known and most popular were those made from 1920–1928 by J. W. Sutcliffe Ltd, a company formed in Leeds in 1885 and happily still making toy boats, as well as oil cans and other pressings. The Sutcliffe Warships were made of heavy gauge tinplate and for this reason, and the relative inefficiency of the P.W.E.s, were much slower than the clockwork and electric boats which have succeeded them.

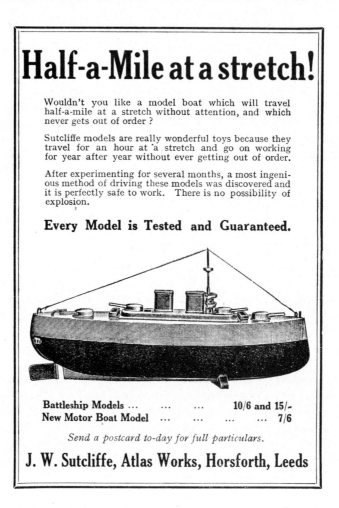

Half-a-Mile at a stretch!

Wouldn't you like a model boat which will travel half-a-mile at a stretch without attention, and which never gets out of order?

Sutcliffe models are really wonderful toys because they travel for an hour at a stretch and go on working for year after year without ever getting out of order.

After experimenting for several months, a most ingenious method of driving these models was discovered and it is perfectly safe to work. There is no possibility of explosion.

Every Model is Tested and Guaranteed.

Battleship Models ...   .   10/6 and 15/-
New Motor Boat Model 7/6

Send a postcard to-day for full particulars.

J. W. Sutcliffe, Atlas Works, Horsforth, Leeds

J. W. Sutcliffe, established in 1885, still makes clockwork and electric boats but the famous P.W.E. Warships were only in production between 1920 and 1928.

One of the later revivals is *Miss England*, a speedboat made about 1948 by Victory Industries Ltd. She is a good example of the genre, with a generous coil of

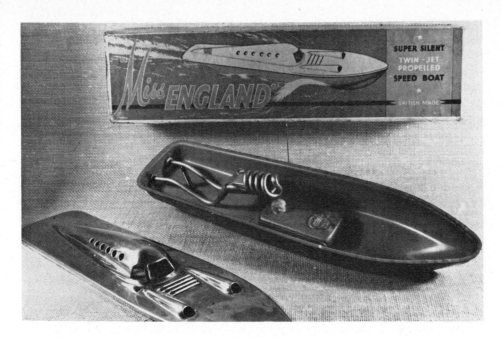

Miss England with the deck removed to show the coiled copper tube which acts as the 'engine'.

large-bore copper tube and is equipped with a rubber filler to prime the tube called a 'supercharger'. With her in the picture (Plate 6) are two Jetcraft launches of about the same date, the largest being some 18″ long. They were made by Jetcraft Ltd of Charlesworth near Manchester.

A popular modification of these simple and silent P.W.E.s was to incorporate a thin metal diaphragm as part of an inboard extension of the tube system. This was stressed like the 'pop' part of an oil can. The varying pulsations now made the diaphragm pop in and out, making a raucous motor-boat noise most satisfying to any normal small boy. These again are believed to have originated in Germany in the late 1920s and they were imported and sold here till the middle of the 1930s. They were called the *Toc-Toc* motor boats, one of which also appears with the fleet in Plate 6 with its flat tin 'driver' which forms the handle of the spirit lamp. Post-war examples were made in Britain—the *Fairylite* by Graham Brothers was about 7″ long; in Germany—the *Puff Paff motorboot* 'under the system Lorenz'; and latterly in Japan. Some of these are also in the photograph. They are charming things and make amusing if rather perilous executive toys for the bath!

Although of great interest, and easily made experimentally in the home workshop, these 'engines' were in no way competitors to the Mamod ME 1 marine engine. Nevertheless, during this first decade after the war commercial competition from other sources was quite severe. The boost given to public morale and to trade by the Festival of Britain in 1951 was reflected in increased demand for all sorts of goods and services, particularly toys and novelties. As in the 1930s Mamod toys were not

47

the only British-made live-steam-engines on the market. Eric Malins estimates that there may well have been as many as thirty firms, many of them ephemeral 'back shed' outfits, making toy engines of various kinds at this time. One or two firms survived until the mid 1960s but the majority were short-lived and had very limited output. Perhaps it is for this reason that the surviving examples of these toys have considerable appeal which is being increasingly recognized today.

There is no British equivalent to Nuremberg as far as the toy industry is concerned, but as might be expected a fair number of steam-toymakers were located in the Birmingham and Midlands area. The traditional metalworking skills and the ease of access to specialist suppliers of materials, and turned, pressed or diecast parts, made (and make) this area a natural choice for those who might wish to market steam toys, but not to manufacture them themselves.

Such a company was the Crescent Toy Co, still active today in Cwmcarn in Wales. They had a brief flirtation with steam in 1947 when they introduced their *Horizontal Engine No 1*. They did not make this themselves and it is not now known where it was made. It was quite a disastrous product. Though extremely cheap, with a simple brass boiler and an engine with a lead flywheel, it had a highly dangerous methylated spirit lamp with the filler tube combined with the handle. This meant that moving the lighted lamp, to put it under the boiler for example, could result in a spirit-covered hand being instantly set alight. Such a dangerous toy could have no future and only about 1,000 were ever made. In 1948, the year after it was introduced, Crescent wisely withdrew it.

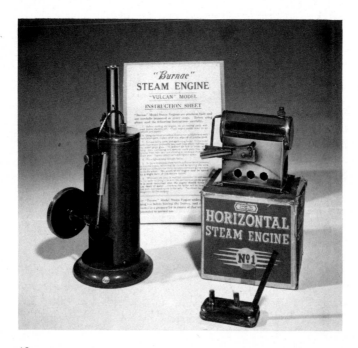

Two post-war rarities; Burnac's *Vulcan* vertical engine and the Crescent Toy Company's ill-fated 'Horizontal Engine No 1'. About 30,000 *Vulcans* were made between 1946 and 1949 by Burnac Ltd of Burslem, Stoke-on-Trent.

The vertical steam-engine, for long a popular type with the German makers, was never produced in large numbers here. Probably the last British engine in this form, and introduced at about this time, was the Burnac *Vulcan*. This was a sturdy and well-made toy (though the box label hopefully says it 'is better than a toy') which boasted a centre flue to the boiler and a speed control achieved (as in the Mamod SE 2) by a throttle device on the exhaust port. The red-painted flywheel and green base were typical of the colours in which most British toys were finished, reflecting standard engineering practice dating back many years into the 19th century. In most engine houses red, green and black were the only paints available, giving a traditional contrast to the polished steel, brass and gunmetal piston-rods and steam fittings. Only Wilesco makes vertical engines in this form any more, which is a pity since they looked well with small toy workshops, or when coupled to one of the more fanciful German water toys or fairground carousels.

Another toymaker, L. Rees and Co Ltd brought out two interesting engines in 1948; one was a fairly simple geared horizontal model equipped with a magnificent brass tubular spirit lamp, which must have cost more than the boiler; the other was a slide-valve fixed-cylinder engine with horizontal boiler and an embossed 'brickwork' chimney as in the best German tradition—a very solid and workmanlike toy that may well have been copied from pre-war Bassett-Lowke designs.

L. Rees & Co's geared horizontal engine with its brass tubular spirit lamp; *Trojan* made by Midland Models of Key Hill, Birmingham which used Signalling Equipment Ltd's die-cast launch unit for the engine; and S.E.L's No 1550 *Major* which made use of thermosetting plastics for the engine mountings.

One of the really ephemeral makers was Desmond Mentha who made the *Rode* horizontal steam engine. So far only one advertisement has been traced for this toy and it appeared in the May 1948 issue of the trade journal *Games and Toys*.

Bailey's Agencies Ltd of London had included in their 1947 catalogue what they described as a 'leading line of steam engines'. The one they illustrated was a typically British design consisting of a twin marine-type vertical engine (oscillating cylinders back to back) set alongside a horizontal boiler with safety-valve, and mounted on a firebox and base of polished aluminium. The extensive use of aluminium was very much a post-war trend and Rees's engines were also mounted on aluminium bases.

Of the many companies active at this time, and listed in Appendix B, one of the most interesting (and in many ways the most mysterious) was the reincarnation of Bowman Models Ltd, now working from an address in Luton, Bedfordshire. The engines were mostly marked with the pre-war trademark of a Red Indian archer, used by the company when in Dereham. The late Mr Bowman Jenkins' family, however, had never heard of the post-war Luton company and have suggested that it may have been run by Captain Smart (of Warboys and Smart) who was a cousin.

A post-war Bowman stationary engine made when the company was located in Luton in Bedfordshire.

Two engines have so far been identified as being of this period. One is a conventional and rather solid, stationary horizontal engine with steam cock and lubricator; indeed certain crudities had led it to be attributed in error to the early 1920s. It is mounted on felt feet and the boiler casing is embossed with the letters 'B.M.'. The other is a much more interesting, totally enclosed, twin-cylinder, marine-type engine with much aluminium in its construction, and again with the embossed 'B.M.' on the casing. It was sold without a boiler and has a most ingenious valve gear. As far as can be ascertained the single-acting pistons are partially rotated, as they move up and down the cylinders, by a form of eccentric crankpins.

50

They thus open and close steam and exhaust ports in the cylinder walls as they move up, down and round at each stroke. One of the author's informants recollects that this engine was shown at one of the post-war British Industries Fairs and a prize of £5 was offered to anyone who could describe how it worked. It isn't known if this was true or, if so, whether the prize was claimed. However, the company apparently got into difficulties and by the middle of 1950 Gamages were advertising their engines at a 50% discount, having presumably bought up the whole lot cheap.

Perhaps the most consistently successful of Malins' competitors was Signalling Equipment Ltd (a subsidiary of J. and L. Randall Ltd, toys and games makers of Potters Bar) who in 1946 introduced a series of four stationary engines under the S.E.L. trademark. The company also made electrical toys, including miniature electric motors and dynamos, though their steam-engines were not powerful enough to drive the generators effectively. The four engines, which with little change were sold for nearly twenty years before being abandoned in 1965, were catalogued as follows:

No 1520 *Minor*
No 1530 *Junior*
No 1540 *Standard*
No 1550 *Major*

They all had the usual (though rather small) oscillating-cylinder engines, the Major with twin cylinders and a whistle. In the early models the engine mountings were parallel to the brass boilers, not at right angles as in the later one shown in the photograph. These mountings have some claim to distinction as being the first (and so far the only) ones for live steam-engines to be made from thermosetting plastic. In 1948 a selection of accessories was added — a circular saw, a grinder and a fan — all with black plastic bodies. They were about 2½" high and were typical of those toys made to be driven by low-powered engines. Not even a hammer head had to be lifted!

Signalling Equipment Ltd was one of the firms which made an attempt to die-cast engines from zinc-based alloys. They made a small launch engine (but not a boiler) by this process. It was sold for home boat-builders and was also used by other manufacturers as a power unit. One of these was another short-lived steam-toy maker, Midland Models of Key Hill, Birmingham, who produced the *Trojan* marine plant. The spirit lamp of this plant was very unhappily devised since the raised tank, having no control valve, ensured that the wicks would flood badly. Perhaps everything getting red hot and producing too much steam was just what the engine needed.

Malins also used the S.E.L. die-cast engine for a time as their ME 3 marine engine, introduced in 1967. This was abandoned about 1972 since it was not very satisfactory and, as designed, needed very great care over lubrication to make it work efficiently.

Before this, in 1957 and 1958, two major changes had taken place involving all the Mamod products. The first was to change the thick flat steel base plates to rather

STEAM ENGINE

MODEL "MINOR" № 1520

REGISTERED SEL TRADE MARK

MADE IN ENGLAND

The smallest in the range of Signalling Equipment Ltd's stationary engines.

lighter and more elegantly shaped pressed bases with corner feet. This meant that the hot parts under the boiler were now clear of the table or whatever was likely to be ruined by having the engine on it and also the general effect was given of a heavier and better quality toy. The other change was the introduction of the vapourizing lamp, which became standard on all engines, both stationary and mobile, until the end of 1976. The earlier lamps with their varying numbers of tubes for cotton or asbestos wicks were more expensive to make, and needed more attention to make them burn satisfactorily, than the perforated metal covered pans that succeeded them. Admittedly some of the fun to be had by using lamp oil, and other heavier fuels, to produce as much smoke and soot as heat, was lost in the change, but some sacrifice has to be made for progress!

By the end of the 1950s the Camden Street works were no longer big enough for the company. It was successfully and steadily increasing its share of the home market and making more than preliminary moves into the overseas markets. Every competitor, even the ephemeral ones, who went out of business or ceased production meant increased turnover. Since the founding of the firm a quarter of a century earlier every move to larger premises had inevitably resulted in a temporary set-back whilst new equipment, machinery and routines were incorporated. And the new premises had in every case, sooner or later, proved inadequate. Furthermore they had all been located within the proverbial stone's throw of the crowded centre of Birmingham, Britain's second city. So plans were now made for a move right away from the congestion of an old city centre which was rapidly being redeveloped. This was to a new site some ten miles away in the heart of the Black Country.

It was here, in 1961, that a one acre site was purchased. Originally it had been a pipe works and when taken over it was still complete with tall chimneys and kilns. The buildings were adquate for present use but, much more important, there was enough land upon which to expand and develop in the years ahead.

1930s

Chapter Five

THE YEARS
OF EXPANSION

THE BLACK COUNTRY is an industrial conurbation, difficult to define with any accuracy, but which lies between Birmingham and Wolverhampton. A cradle of the Industrial Revolution it had for centuries been a major source of coal and iron. It was a landscape of rounded hills patterned with railways and canals, factories and farmsteads, rolling mills and brickworks. Here was a great concentration of foundries and blast furnaces with their perpetually smoking chimneys 'Black by day and Red by Night'; to many the very breath of Victorian industry. Most of the old towns and villages within the conurbation had (and often retain) their own character and individual industries — wrought iron chains and ships' anchors in Netherton and Cradley Heath; scythes and nails in Belbroughton; locks at Willenhall and harness and leather work in Walsall.

Today the coal and iron is largely worked out and the older heavy industries are giving way to cleaner and lighter manufactures, of which precision engineering is one. Into this the Malins enterprise fits well. The new works are situated at Quarry Bank, Brierley Hill, an old centre of mining and ironmaking. The site and existing buildings were purchased in 1961 and the move from Camden Street was made in the following year — as Eric Malins puts it: 'We moved between the steam roller and the traction engine'.

For it was in 1961 that Eric Malins designed and introduced the SR1 steam roller. This was an important development in the company's products since it marked a move away from an exclusive reliance on stationary engines to those much more exciting ones which run about under their own steam. The roller was an immediate success. There had only been some German ones available hitherto, with the exception of a poorly produced English one sold by Gamages about 1960. Indeed it was the inadequacy of this that spurred Eric Malins to go ahead with his own design. This was based on the Minor 2 non-reversing engine, mounted on a firebox, with a spring belt-drive from a pulley on the flywheel end of the crankshaft to the nearside rear wheel or roll. The front and rear rolls were made from aluminium pressure die-castings and had polished rims. These were subsequently slightly modified to the present zinc-alloy type, and the engine itself was changed to incorporate the reversing block (really essential, after all, for a steam-roller) in 1967. The modern spring re-set whistle and the use of pop rivets instead of the earlier nuts and bolts date from this time too.

The success of the roller encouraged the further and predictable modifications of the design to produce a traction-engine. The firebox, boiler and engine were standardized for both, but proper wheels were vital to the effective appearance of the new model. English traction-engine wheels were built up from castings and wrought iron spokes rivetted to fabricated iron rims. They were complex, expensive and very characteristic. If a toy version did not have wheels that looked right it would not sell and, if such wheels had to be made in anything approaching the traditional way they would be so expensive as to price the toy out of the market. The solution was ingenious tools, which enabled a very close copy of the traditional wheel to be made on the die-casting machine with great expedition and, at the same time, incorporating very fine detail of such things as rivet heads. So, with a new smokebox to carry a simplified perch bracket for the front wheels, the TE1 traction-engine was introduced in 1963, (Plate 9). The two year delay after the roller was, of course, in part due to the move to Quarry Bank.

Soon the traction-engine established itself as the most popular toy that the company had ever produced. Again, as with the roller, some changes have been made over the years, notably the reversing engine and the spring whistle, both dating from 1967. The SR1 and TE1 then became the SR1a and TE1a. Both the roller and

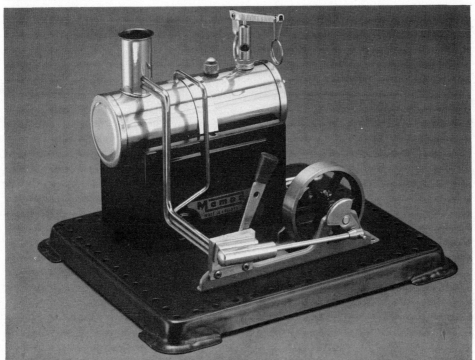

The popular Mamod SE 2(a) stationary engine.

the traction-engine are still in very active production and on one day in 1974, when the company was visited by a reporter from that very respected journal *The Engineer,* six hundred traction engines were turned out. In total over 350,000 have been sold.

Accessories — working toys — for the engines have never figured very largely in the Mamod programme, but two important new ones were introduced in 1968 for use with the new mobile engines. These were the lumber wagon and the open wagon. Both are characteristically sturdy products using the standard die-cast aluminium wheels fitted to the front of the traction-engine and common underframes. Simplification and standardization are keynotes of the company's attitude to production and, apropos some of his erstwhile competitors the most characteristic thing that Eric Malins says is: 'They didn't get their manufacturing economics right'.

Although Mamod stationary engines had had their base plates drilled for building into Meccano models (as had the Bowman engines before them) Meccano Ltd had produced their own engine in 1929 as we saw in Chapter Two. This was an attractive little toy but it was not very popular and manufacture had ceased by 1935. However, thirty years later the company decided to re-introduce a steam-engine and asked Malins to make one. Meccano Ltd laid down the main design characteristics of the base and side plates for the crankshaft and layshafts. These were identical with the 1929 model. Eric Malins then contributed by incorporating a horizontal boiler and designing a simple reversing block to be used with the standard oscillating engine. This has resulted in a popular and effective engine to take its place alongside the more mundane clockwork and electric motors in the Meccano range.

A similar reversing block was also soon used on the Mamod engines and became standard on the two mobiles — the roller and the traction engine. It is interesting how

A rare example of the 1929 Meccano steam-engine owned by Geoff Wright of Henley-on-Thames. Fully restored by Brian Wright and Roy Hallsworth it makes an interesting comparison with the contemporary Mamod made engine on the right.

A simple steam trip-hammer incorporating the modern Meccano engine.

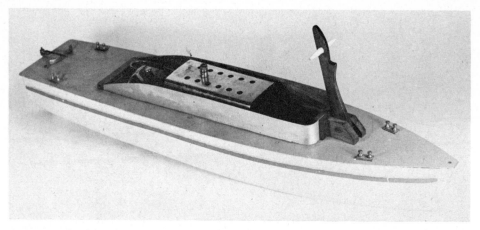

The Perico Company's launch made in the late 1960s and using the Mamod ME 1 steam engine as its power unit.

small things can pinpoint dates of developments in companies. Stephen Malins, Eric's youngest son, who was still at school at the time, remembers clearly that the Meccano engine was introduced in August 1965 because the family was then on holiday in the Isle of Wight and they received an engine dispatched from the works by post to test its prototype packaging. It apparently arrived undamaged and the packaging was duly approved.

The decade of the sixties was one of steady expansion. The steam-toy business, even in the heyday of German domination in Edwardian times, was never much subject to the whims of fashion and, with the expense and complications inseparable from the introduction of too many new designs, it was a steady concentration on improved production methods that brought the best results at this time. Geoffrey Malins, now in his seventies, continued as Chairman though practically the full responsibility was falling upon Eric as managing director. In 1968 *his* son Stephen left Handsworth Technical School with little reluctance and joined the family firm. This was a time of new developments and ideas in sales and marketing, with increasing concentration on exports. English competition was virtually at an end and, apart from the model-making companies, like Stuart Turner Ltd, who still make one or two 'toys', Malins was now the only surviving live steam-toymaker in Britain.

There followed a year or so of consolidation, during which time Stephen, as well as learning the business, was conjuring up new variations on the basic theme of boiler, lamp, oscillating reversible engine and (preferably) four wheels. At times the prospect of creating anything very much different from the traction-engine, without the addition of an uneconomic range of new components, must have seemed daunting, but by the summer of 1972 a promising looking prototype of a steam wagon was on his office desk. This went through the usual development and productionizing stages in a remarkably short time and the first models were in the shops by Christmas—a bare twelve months after the original conception.

Mr Eric Malins, Managing Director, with his son Stephen who was largely responsible for the design and production of the steam Wagon first introduced in 1972.

Plate 13 shows this attractive toy based on the Foden steam wagons with their overtype engines, which were still to be seen on the roads of Britain when Malins was founded in the 1930s. Many components of the Mamod wagon are standard; for example the whole of the front, including the engine, is virtually identical with the traction-engine, with the addition of a chassis built from two long channel sections rivetted to the firebox. The belt drive from the crankshaft is taken to a countershaft and thence via another spring belt to the rear wheels. This gives a slower and more realistic action than a direct drive would provide. The rear wheels are die-cast in zinc alloy and are most faithful copies of Foden prototypes, once again cleverly giving the appropriate character to a period vehicle. The cab and wagon body are heavy steel pressings and are easily removable for attending to the engine before a run.

The steam wagon was a peculiarly British vehicle and as far as can be discovered the Mamod wagon is the very first live steam toy version to be produced. The early German makers made a few steam commercial vehicles, but none approaching the charm and period flavour of a Foden wagon. Moreover it is of quite massive construction, far removed from any suggestion of tinplate.

With this addition to the range, expansion of the company soon became essential and, having space available on their own site, plans were made to extend the works to a total of something like 25,000 square feet. The output was going up as was the number of people employed and in the mid-1970s this rose above the hundred mark for the first time in the firm's history. Some work, typically soldering, used to be done as 'outwork', that is to say in the operatives' own homes, but all this had been brought into the factory by this time. The new methods and new tooling will be described later but another major change was incorporated in the extensions in the

Malins (Engineers) Ltd works in Thorns Road, Brierley Hill.

form of a new office block. Hitherto the facilities for both management, office staff and visitors had been of the simplest, crowded and inconvenient. The new offices and extensions were opened (with an uncharacteristic burst of publicity in the *Birmingham Post*) in the middle of 1975. Now for the first time there is a proper showroom in the best Nuremberg tradition and comfortable, well carpeted offices. Stephen Malins has a case containing a representative collection of the company's earlier models in his office—few firms in the past were very good at preserving their own history—together with some attractive Lowry prints on the wall. The original painting used in the brochure for their latest toy faces his desk.

Mr Geoffrey Malins had continued as Chairman during all the planning and new building work. The author remembers the last time he met him, in February of 1975, a very elderly man limping round the works with a stick. We were looking at an early example of a Bassett-Lowke locomotive, kept in the development department. 'They made quality' Mr Malins growled, then—'We make quantity as well.' Unhappily he did not live to see the opening of the new works. He died, aged eighty-three on the 12th of May, having been at the helm of the company for over forty years.

The artistic and, even more important, the commercial success of the steam wagon encouraged the company, with Stephen's active participation, to look to further development on similar lines. The result of this work has been the resoundingly successful Steam Roadster introduced in 1976 (Plate 8). This is a logical development of the wagon story, firstly because it is a solid well-built toy, over 15" long and weighing 4 lbs, and secondly because it has a strong period flavour and is quite consciously designed to appeal as much to parental nostalgia as it is to present day youngsters. This is not an exclusive trait of Malins, since many toymakers have been delightfully old-fashioned in their designs, often basing their toys on prototypes twenty or more years earlier.

The Roadster uses the standard oscillating-cylinder engine with reverse, belt transmission similar to the wagon and virtually standard boiler and lamp. In this case the boiler backhead (to use a locomotive term) is secured to the scuttle of the car with pop rivets. All these components have been very cleverly incorporated into a chassis looking much like a sporty pre-1914 two-seater open roadster. The flywheel is by the driver on the off side and is the same as that used on the Meccano engine. The die-cast artillery-type wheels with PVC tyres, the substantial steel body pressings and running boards together with the twin brass plated headlamps on the scuttle are all produced on the standard machines used for the components of other products.

It is all imaginative and economical. The exhaust steam issues from the side; a conventional car-type exhaust pipe is not fitted. To reduce the need for the lubrication that such toys often don't get nylon bushes are used for the bearings of the crankshaft and transmission. The front suspension is by vertical coil springs and a modified 'Ackerman' steering system (a bit fierce!) is fitted. It is the first proper working steam car that has appeared for over half a century.

What next? Well, always the consciousness of the need for greater safety. Increasingly repressive legislation has led to a review of the open methylated spirit lamps. Though they have been used without disaster for generations, the solid fuel of the Meta or Esbit type is so much more convenient and inherently safer than an open bottle of spirit, that since the beginning of 1977 all the Mamod engines have been supplied with solid fuel heaters — the fuel itself being made by the company.

Then there are the next products. What will they be? A newly designed range of stationary engines is already well advanced and another boat is always possible. After all, Stephen Malins with his complete lack of experience in the Navy might well succeed where his father and grandfather failed!

1930s

Chapter Six

MAKING
MAMOD ENGINES

TOWARDS THE END of the 18th century the great Dr Samuel Johnson paid a visit to the Soho Works of James Watt and Matthew Boulton on the outskirts of Birmingham. He later wrote that Mr Boulton 'with great civility led us through his shops. I could not distinctly see his enginery. Twelve dozen buttons for three shillings. Spoons struck at once'. Dr Johnson was neither the first nor the last visitor to a factory to be unable to see (or understand) the 'enginery' or to be impressed with the wonders of quantity production. Happily Malins do not make spoons or buttons but a latter day Dr Johnson might still record with some astonishment 'traction-engine wheels struck at once' when seeing the modern die-casting machine which makes them at Brierley Hill.

Here is a proper engineering works organized to produce miniature engines. Today a steadily increasing proportion of the components are being produced in-house and reliance upon outside sources, other than for raw materials, is less than it was in the past. The starting point is the stores, from which the shops are supplied with either finished (or semi-finished) components made by outside suppliers, or with raw materials, such as sheet steel, copper tubing and zinc alloy. These stores run like a backbone down the works. The machine shops are on one side; the assembly departments on the other. There is a computor system for the stock control of raw materials and components which are issued according to the schedules for batch production.

The die-casting machine, an automatic EMB No 8, is a good starting point for our tour. Here, ingots of zinc alloy are fed into the melting pot and, depending upon which of a large range of tools is in the machine, a steady stream of components emerges—chimneys and smokeboxes for traction-engines and rollers, bases for the accessories, lamps and steering-wheels for the Roadster and, of course, all the varied types of wheels required, such as flywheels, pulleys and the beautifully detailed wagon and traction-engine road wheels. Having been cast the components are then 'clipped', i.e. cleaned up by having runners and any flash removed on power presses. They are then ready for painting or plating. The brass-plated headlamps for instance are an attractive feature of the Roadster.

The drilling section in the machine shop.

An electrostatic paint-spraying line is installed with gas-fired baking ovens. Girls put the various components onto appropriate carriers, which then move through the degreasing bath, past automatic spraying heads and through the stoving sections on a continuous track, so that they come out at a station near the loading point, there to be removed for further finishing. Most wheels and rollers, which are automatically painted all over, now go to lathes to have the rims skimmed to remove the paint and leave them bright; afterwards the wheels are lacquered all over by hand.

Meanwhile, in the power press section steel sheet is blanked, pierced and raised for some of the smaller base plates, fireboxes etc. Larger components such as the bodies and chassis for the Roadster and the wagon are still made on more powerful presses by outside suppliers.

There is a very large hand press section in which a very wide variety of punching, pressing and bending operations are done. A good example is the making of the boilers. These come in a variety of sizes and are all made of brass. They are made by outside suppliers as deep drawn 'cups' with one boiler end integral with the shells. Holes for filler/safety-valve bushes, level plug bushes and fixing rivets are punched out on the fly-presses. The bushes themselves are inserted from inside, peened over and soldered in place in a gas-fired oven. The flanged boiler ends are also made in this section and are soldered on at the same time. Other operations include the

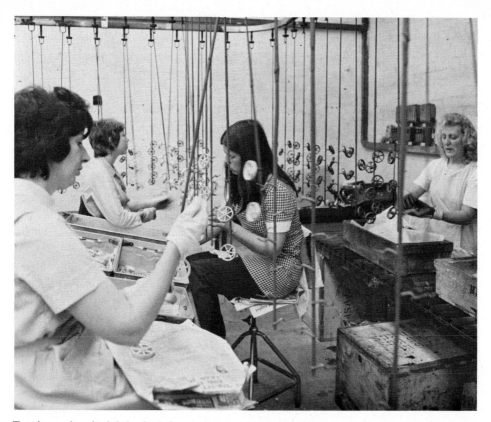

Traction-engine wheels being loaded onto fixtures which carry them through the automatic spray-painting plant where they emerge (in the background) to be removed.

bending of the steam and exhaust pipes to shape and such items as the canopy supports for the traction-engines. Pistons and big-ends are also assembled here by pressing them onto the steel piston rods. In so doing the pistons are slightly distorted, so a further interesting operation is 'sizing' them by forcing them through a precision die to shave off just enough metal to make them fit the cylinders accurately.

The brass cylinders themselves are bought out from I.M.I. Kynoch works and for many years deepdrawn on cartridge-making machines. The port blocks are made from brass drawn sections cut to length and soldered onto the cylinders. This is done in a short gas-fired oven. The port faces are ground flat on linishing machines, the latest being a fully automatic one, and of course, the fixed port faces and, if fitted, the reversing blocks are also linished to a fine, flat finish to make sure that they are steam tight. The cranks are stamped out of mild steel and the crankpins located and stud-welded on. They are then sent out to be zinc-plated. The boilers and basic engine components now being finished, the brass work is polished and hand lacquered to retain their bright and attractive finish. They now come together with other items for final assembly.

65

Wheel turning and the power press section.

Over the years an increasing use has been made of fast-fixing 'pop' rivets for fastening parts together. In Germany the old tab and slot principle was much in favour. Here nuts and bolts long held sway but today something quicker and in many ways more reliable is needed, and here the 'pop' rivet comes into its own, particularly since copper ones are self-sealing and can be used directly in the boilers. All brass fittings are made out, the vast majority by Brass Turned Parts Ltd, an old established Midlands firm which has been doing this work for Malins for over thirty years. These fittings include safety-valves, whistles, level plugs and pulleys for the cars, wagons and stationary engines. Most of the sub-contract work reflects the company's long connections with suppliers in Birmingham and the Black Country.

One of the key operations in the assembly shop is hand soldering the copper steam and exhaust pipes to engines and boilers. A steam-tight joint is an obvious requirement, but it is equally important not to leave pipes blocked up with surplus solder in the wrong place. So, to check that this has not happened and to make sure that everything is properly assembled all engines are run on compressed air at this stage. The boilers are air tested for leaks and general soundness at pressures up to 75 or 80 pounds per square inch. Then the engines are run at their normal working pressure of between 10 and 15 p.s.i. For this test they are lubricated with a light oil, since they are run cold. As everyone knows who has driven a steam-engine, toy,

Steam and exhaust pipes being soldered to a wagon in the assembly shop.

model or full-sized, an appropriate motor or steam oil should always be used when actually being steamed. The safety-valves are set to 'blow off' at 1½ times the working pressure of the engine — many times lower than the safe working pressure of the boilers.

After these tests the mobile engines get their wheels fitted together with appropriate driving belts and the final finishing touches are put on—transfers and the like. They then move on to the packing department to be put into their attractively illustrated boxes complete with filler funnels and now, packs of solid fuel tablets as well. And well under 1% of them will ever be returned for repair.

Today a fair amount of chromium-plated steel strip (with blue plastic protective coating) is punched and formed into such items as the traction-engine fireboxes, the marine engine boiler casing and the bonnet of the Roadster. Only when all is finally assembled is the protective coating removed. Another recent change has been the extended use of plastics material, which was previously confined to heat-resisting handles on reverse levers and spirit lamps. Now the Roadster has required ribbed p.v.c. moulded tyres and steering-wheel rim and a black plastic moulding to represent the leather-upholstered seats. The steering-wheel is attached to its column by 'Loctite' and nylon bushes are used for the wheel and countershaft bearings.

Front axle assembly being fitted to a traction engine.

Finishing touches being put on a traction-engine after final test.

There is nevertheless a very real limit to the use of plastics materials in a live-steam engine and *real* steel, brass and copper will continue to be the staple materials of the steam-toymakers. Fortunately this use of expensive traditional materials also adds tremendously to the effective appearance and appeal of these engines.

Chapter Seven

STEAM TOYS
INTERNATIONAL

EVER SINCE MAMOD toys first appeared over forty years ago the company has been actively concerned with the effective marketing of them. Alongside the traditional advertisements in the popular boys' papers, with the usual special emphasis for Christmas, the firm has regularly exhibited at the trade fairs. This started in the 1930s when the annual British Industries Fairs, then in two sections in London and Birmingham, provided a general showplace for everything from heavy engineering to domestic wares. Mamod engines were shown at most of these until the B.I.F. came to an end in 1955. But it was, and still is, at the large international toy trade fairs that the greatest impact is made. The Brighton Fair, for many years a January Mecca for the trade, moved in 1977 to the new National Exhibition Centre near Birmingham. Here it was re-titled the British Toy and Hobby Fair and, almost on their own doorstep, Malins had their largest and best designed stand ever, putting considerable emphasis on wooing the foreign buyers. Their publicity has for some time been printed in German, French and Spanish as well as English.

Since 1968 they have exhibited at the even bigger Nuremberg Toy Fair, though their engines have not been acceptable to the Germans themselves owing to a legal/technical difficulty widely believed to have been created deliberately. This is simply that there is no water level gauge on the boilers. The German toys have what might be considered a dangerous (and expensive) glass disc bolted onto the ends of their boilers providing a curious and very 'un-engineering' see-through effect until the glass gets obscured by deposits of lime etc. The battle is not yet resolved but Malins have now developed what is hoped to be an acceptable gauge for little greater cost than the existing level plug which it would replace.

At home Malins sell direct to the retailers — some 2,000 of them. Their first representative was appointed in 1970 and today they have four salesmen covering England and Wales. Previously they had employed agents, as they still do in overseas countries. The export trade really took off in the mid 1960s with the great popularity of the rollers and traction-engines. The major countries to which they are sold are America, Australia and Canada, with an increasing interest being displayed by the French. Something like 25% of their output now goes overseas and this proportion is rising. Just before the end of 1976 their U.S.A. distributor rather underestimated demand for the new roadster. For the first time in the company's history they turned to air freight as a means of getting an extra 1,100 steam cars into the American shops in time for Christmas. It was the largest consignment of toys the air freight people had ever handled and hopefully resulted in an extra thousand or so American children enjoying a steam-powered Christmas.

Another country to which the engines are exported is Sweden. Here, however, liquid methylated spirits is apparently not available, so since 1970 special trays to take the solid fuel in tablet form have been made to be interchangeable with the standard lamps. It is these former specials which are now standard for both home and export models.

The company has long been a member of the British Toy Manufacturers Association and its sister organization, the National Association of Toy Retailers who have for some time made annual awards to the makers of the most meritorious toys. In 1969 and again in 1975 Malins were given one of the very few and much coveted Council Awards. In 1972 the 'top' Toy of the Year award was given by general consent of the trade to the Mamod Traction Engine.

The world has shrunk considerably since the 1930s and the European Common Market has seemingly made our corner of it smaller than ever. Although Mamod engines are the only live-steam toys made in the U.K. the international field still holds one or two rivals—in fact there are three. One is in U.S.A., one in Sweden and perhaps the most important, in Germany. Although our main concern in this book is with the British contribution to the story of toyshop steam-engines it seems appropriate to give some consideration to the few currently available foreign examples. There is evidence that the number of makers may soon increase to match the increasing popularity of these steam toys. They are widely being recognized as being so much more interesting and demanding than those mere wind-up-and-watch or just switch-on playthings.

So let us start with America. The Jensen Manufacturing Co Inc. of Jeannette, Pennsylvania was established by T. H. Jensen Senior in the 1930s and makes a range of stationary engines, no mobiles, very much in the German tradition, with embossed imitation brickwork boiler casings and chimneys, some with handrails and ladders 'for extra realism'. All but one have horizontal boilers and there is one vertical engine, but the vertical boiler steams a horizontal engine and, conversely, the vertical engine (a double-acting oscillating-cylinder model) is teamed up with a horizontal boiler. One of the engines is supplied as a kit for home assembly, mostly on the old tab-and-slot principle and with the engine bracket needing to be riveted to the baseplate. The boiler has a large round glass window in the end, much like the Wilesco engines, as have a number of other models in the catalogue. Altogether thirteen models are made, mostly with lever reversing slide-valve engines and imitation governors. Ten of them are equipped with electric immersion heaters in the boilers. The three smaller ones are fired with solid dry fuel. A range of workshop accessories is also available for building into machine shops. These engines, and the accessories which include a dynamo, are heavily and solidly built and are advertised as being of great educational value in the teaching of physics. Perhaps for this reason

A typical stationary engine and generator made in the U.S.A. by the Jensen Manufacturing Company of Jeannette, Pennsylvania.

the larger models are consciously aimed at schools and educational institutions. The company does not, however, appear to be interested in selling them very actively in Europe and very few have been seen in Britain.

A newcomer to the live-steam scene is the Swedish company of Ab Alga of Vittsjo. Some years ago a small vertical marine engine was available in Britain made by Eskader of Stockholm, but this does not seem to have survived. Ab Alga's engine, named the *John Ericsson* is new both in production and in appearance. The power unit is a fairly conventional oscillating-cylinder engine with a combined speed control and reversing lever. It is the boiler that has been given a most unusual and attractive form. Functionally it is a rather small vertical pot boiler but it is cleverly incorporated in a polished brass and black 'haystack' casing with a surrounding handrail and is capped with filler, whistle (much like Mamod's) and a simple weighted lever-type safety-valve. Both boiler and engine unit are mounted on a heavy pressed steel base, drilled as always these days for Meccano and other construction sets. It is supported on four plastic sucker feet to prevent damage to tables and to stop it vibrating its way onto the floor. The whole is a most curious mixture of tradition (the early type boiler, the toyshop engine) and modern Scandinavian design reflected in the shape and form of the smokestack and the advanced safety features. Ab Alga also make a transmission (shafting) with a lathe and a polishing machine for the engine to drive. Unfortunately these are very poor in design and not in the same class as the engine. However, as their leaflet puts it, 'here

The *John Ericsson* engine from Sweden with the haystack type boiler casing and oscillating-cylinder engine. It is made in Vittsjo, by Ab Alga.

is a steam-engine that satisfies today's requirements of safety and through its somewhat unconventional look in a favourable way differs from all others'. As indeed it does.

In Germany the metalware company of Wilhelm Schröder and Co. of Ludenscheid makes a number of steam toys well known both there and abroad under the *Wilesco* trade mark. There is some doubt about the length of time they have been making steam-engines but they were not among the well known early makers such as Bing, Marklin, Plank or Falk, or even Fleischmann who were still making some engines as late as 1965.

The Wilesco range is an imaginative one and comprises stationary engines with both fixed and oscillating cylinders, a steam-roller, a traction-engine and a full range of accessories. The stationary engines are conventional enough, some fired with Esbit or similar solid fuel tablets, others fired with electric immersion heaters in the boilers. The traditional embossed tinplate brickwork of the boiler casings and chimneys has been given a curious 'Wilesco character' by being enamelled in a copper colour. About 1958 they brought out a 'Working model of an atomic power station', an extraordinary toy which 'takes the conventional toy steam-engine a stage further' as their publicity said. The power unit was a standard horizontal piston-valve mill-engine but the boiler was disguised as a spherical reactor with simulated cooling towers. The boiler was heated by electricity and 'the switches are coupled to the regulating rods'. Its method of operation is not clear to the author,

73

The D40 traction engine.

Steamship boiler and engine.

Two models from the current Wilesco range.

since it was stated that the 'supply water tank is sited within the cooling tower where water is pressed into the tower for the exchange of heat by means of the pressure of steam'. It was not popular however and examples were being sold off cheap in 1960 by Gamages at £5 19s 6d each (£5.97). Perhaps some cautious parents thought it was radio-active.

The current steam-roller is an attractive model, called in all countries *Old Smoky*. It has a fixed cylinder and proper chain-steering via a handwheel. The traction-engine is visually less satisfactory, having the same perch bracket as the roller with a wholly uncharacteristically mounted pair of wheels underneath. Apart from a rather longer canopy it is otherwise the same as the roller. All engines except two very cheap export only stationary ones have the large glass discs mounted in the ends of the boilers.

The accessories are the most numerous of any of the surviving manufacturers and are clearly in the old German tin-toy tradition. There are the usual machine-tools of course; drill, saws, press and grinder, but also a drop-hammer, a triple stamp mill, a flexible shaft drill or grinder, a planer and a guillotine. Outside the machine-shop there is a novelty colour changing wheel, a cement mixer and, of course, a dynamo. So with the transmissions that are available quite elaborate set-piece toy workshops can be devised and driven by the larger engines. Unfortunately none of the tools have any of the little tin men operating them that were such a pleasant feature of the earlier German makers.

From whichever country it comes the live-steam engine has unique qualities not even remotely present in any other toy. It is a true heat engine deriving its motive power from the energy contained in the fuel—methylated spirit or its modern solid equivalent. This heat energy is converted into mechanical power in exactly the same way as in the prototype reciprocating engines now alas mostly 'preserved'. Should the engine be coupled to a dynamo, then other forms of energy can also be created. Electro-magnetic induction, discovered in the early years of the 19th century by Michael Faraday, will produce a current of electricity from the coils and magnets of a toy generator just as it will from those of a turbo-alternator in a nuclear power station. This can light a flashlamp bulb showing incandescence, heat and light, in the tiny wire filament. What other toy can demonstrate so complete a chain of energy conversions on a table top?

Again the steam toy has the charm and interest of a living thing. It is not wholly predictable, it needs care and attention to get the best out of it. It carries with it the spice of apparent danger; it fizzes, it smells; it makes noises and it's hot. In short it is a demanding toy and as such gives just that much more in return. And should it be a mobile toy—a car, a steam-roller or a traction-engine—so much the greater scope for imaginative play and perhaps the steamboat, which makes its journeys out of reach of the young mariner's direct control, offers the greatest stimulus.

There is another aspect of the steam-driven toy which has been implicit in our whole approach in this book and that is the interest of the collector. Since comparatively few modern engines are available today the period charm of the really old ones is being increasingly appreciated. The collectors of early toys (and they are legion) have not yet given full attention to the claims of British steam toys in particular. The engines lack perhaps the instant glossy appeal of the gaudy lead and tinplate products of a pre-war Germany but they have much solid merit to recommend them. The majority, when in reasonably good condition, can be safely steamed for our enjoyment even when half a century old. Should repairs be needed their sturdy qualities, unit construction and ease of assembly make renovations reasonably straightforward. So the collectors, preserving the best of the past in the interests of the future, have a real, albeit perhaps a minor, role to play in keeping alive the excitement and mystery of steam engineering.

'Be a *real* engineer' said Bowman's advertisements in the 1930s. Real engineers today are making their contributions to our wealth and comfort in fields far beyond the simple steam-engine. But the steam age, which powered Britain's Industrial Revolution, can still be savoured, writ small though it be, on our kitchen tables and playroom floors through the surviving examples of generations of toyshop engines.

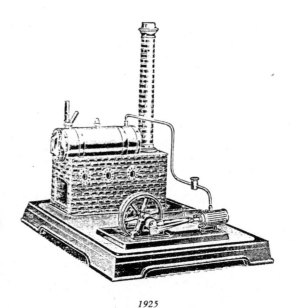

1925

MAMOD PRODUCTS
1934-1977

STATIONARY STEAM ENGINES

ALL HAVE BEEN horizontal engines; methylated spirit fired first with wick lamps until about 1958, then with vaporizing lamps until 1976. From 1977 solid fuel has been used. Flat steel bases were in use until about 1957 when the change was made to pressed steel.

MINOR 1

The smallest in the range. An overtype engine first made c. 1937 and revived after the war. The crank-disc acted as a flywheel until a spoked flywheel was introduced in 1954.
Still in production.

MINOR 2

A horizontally-opposed twin-cylindered engine was first made for a year or so about 1935/6. The present single cylinder overtype version was introduced in 1949.
Still in production.

S.E.1

Was first made for Hobbies Ltd in 1936 and revived in 1946 with a solid crank-disc flywheel. Spoked flywheels and longer engine frames were introduced from 1954. In 1966 became the:

S.E.1a

which had minor changes only, notably the omission of screwed unions.
Still in production.

S.E.2

Was first made for Hobbies Ltd in 1936 and revived in 1947 with solid crank-disc flywheel, exhaust throttle for speed control and geared countershaft. Spoked flywheel and longer engine frame introduced and gears abandoned in 1954. In 1966 became:

S.E.2a

with the addition of a reversing block on the engine and the loss of the throttle. Spring re-set whistle in 1968.
Still in production.

S.E.3

When introduced in 1936 for Hobbies Ltd this was a larger version of the S.E.2. It was not made after 1939. The new S.E.3 was introduced in 1957 in its present form, ie: a superheated uni-directional twin-cylinder engine. The spring re-set whistle dates from 1968.
Still in production.

S.E.4

First made about 1935 for Hobbies Ltd. This was the largest model with geared countershaft. What may be the prototype is still in Malins' possession. It was not made after 1939.

MECCANO STEAM-ENGINE

Made for Meccano Ltd, from 1965. Reversing block also incorporated in other models.
Still in production.

MARINE ENGINES AND BOATS

M.E.1 MARINE ENGINE

First made for Hobbies Ltd about 1937 without a boiler casing. Revived after the war in 1949 for fitting in the *Meteor* yacht. Inclined engine and integral propeller and shaft. Solid fuel tablets from 1977.
Still in production.

M.E.2 MARINE ENGINE

Made from 1956 until 1967. This was similar to the M.E.1 but with a vertical engine and no propeller or shaft.

M.E.3 MARINE ENGINE

This was a similar unit to the M.E.2 but was fitted with a die-cast engine made by Signalling Equipment Ltd. (See a related plant *Trojan* with the same engine, page 49). In production from 1967 to 1972.

METEOR STEAM YACHT

First made in 1949. Steel hull with M.E.1 engine. A single screw boat measuring 25" long. Not commercially successful and abandoned after 1952 after some 1500 had been made.

CONQUEROR ELECTRIC LAUNCH

This was identical in outward appearance to the *Meteor* but was fitted with a battery powered electric motor. From 1949 to 1952.

MOBILE STEAM TOYS

All were issued with vaporizing methylated-spirit lamps until the beginning of 1977 when solid fuel tablets were substituted.

S.R.1

The steam-roller was first made in 1961 without a reversing engine and with red painted centres to the rolls. In 1967 it became —

S.R.1a

when the reverse block was added, the throttle abandoned and the rolls changed to the present types. Spring re-set whistle 1968.
Still in production.

T.E.1

The traction-engine was introduced in 1963 and was substantially the same as the present —

T.E.1a

which succeeded it in 1967 when the reversing block was added and the exhaust throttle abandoned. Spring re-set whistle 1968.
Still in production.

S.W.1

The steam wagon was introduced in 1972. It was painted red and green until 1976 after which the green was changed to blue. (All blue wagons use solid fuel.)
Still in production.

STEAM ROADSTER

First introduced in 1976.
Still in production.

ACCESSORIES AND WORKING MODELS

BRASS PROPELLERS and
BALL-BEARING TURNTABLES

These were among the company's first products made from 1934–1937 and sold under the name 'G. M. Patents'.

LINESHAFT C.1 and
LINESHAFT C.2

These were introduced about 1937 and were first made with cast-iron pedestals on Meccano drilled steel bases. The C.1 was dropped from the catalogue about 1950. The larger one, now fitted with five pulleys, is *still in production.*

M.1 POWER HAMMER and
M.2 POLISHING SPINDLE and

M.3 POWER PRESS

These three models were first made about 1937 and were revived soon after the war. At first they were made of cast iron but about 1950 this was changed to a zinc based alloy, Mazak. The lineshaft pedestals were also made in Mazak from this time.
Still in production.

THE GRINDING MACHINE

This was added later (the pillar is the same as the polisher) and hardly any variations have taken place in these four models over forty years.

L.W.1 LUMBER WAGON and

O.W.1 OPEN WAGON

These two trailers were first produced in 1968 as adjuncts to the Roller (1961) and to the Traction-engine (1963).
Still in production.

Appendix B

STEAM-TOYMAKERS IN BRITAIN

ALTHOUGH NO SPECIAL claim is made that this is complete it is a serious first attempt at a comprehensive listing of British companies, large and small, which have made steam toys and accessories from Victorian times to the present day.

COMPANY	DATES MAKING STEAM TOYS	NOTES
Bailey's Agencies Ltd, London	1948	Twin-cylinder vertical oscillating-cylinder engine with horizontal boiler.
Bar-Knight Model Engineering Co. Ltd, Glasgow	1920– 1925	Horizontal mill engines. Vertical oscillating engines. 'O' gauge oscillating-cylinder 0-4-0 locomotive.
W. J. Bassett-Lowke Ltd, Northampton	1901– 1950s	Locos, stationary and marine engines. Mostly scale models but some were toyshop engines.
J. Bateman and Co., London (Est 1774)	c. 1880– 1890	Good quality dribblers & other brass engines. Also known as 'The Original Model Dockyard'.
Bedington Liddyatt & Co., London	1919	A vertical engine with oscillating cylinder, spun metal base and chimney top.
Bond's o' Euston Road Ltd, London	1880s to present	Mainly castings and materials; included some toy locos, marine and other engines in early days.
Bowman Models Ltd, Dereham	1923– 1935	Stationary engines, locos and steam launches with accessories.
Bowman Models Ltd, Luton, Bedfordshire	1947– 1950	A company using the Dereham trade mark but not under Bowman Jenkins' direction. Only two stationary engines so far known.
British Modelling and Electrical Co., (later British Engineering & Electrical Co., Macclesfield	1880s– 1920s	Dribbler locomotives and parts for home constructors.
Burnac Ltd, Burslem, Stoke-on-Trent	1946– 1949	Only one vertical engine was made.
Clyde Model Dockyard, Glasgow (Est 1789)	c. 1880s –1930s	Gauge 'O' locomotives, stationary engines in early days and an outboard marine engine.

*Crescent Toys Ltd., Cwmcarn	1947	A cheap brass horizontal engine. Only one marketed.
Graham Bros, London	c. 1948	The *Fairylite* P.W.E. launch.
Hardy Bros., Hove	1950	Single, twin or three cylinder piston valve marine plant.
*Hobbies Ltd., Dereham	1923–1938	Designs and kits for steam launches though engines bearing their name were made by Bowman or Malins.
W. H. Hull & Son, Birmingham	1884–1930s	Early brass engines and later 'O' gauge locomotives.
Jetcraft Ltd., Charlesworth, Manchester	c. 1946–1950s	Small range of good tinplate P.W.E. launches.
W. Joanes & Co., Stoke Newington	Mid 1920s	A range of workshop toys but no steam-engine has been recorded.
W. H. Jubb, Sheffield	1915–1922	Cheap steam-engines and some scale model locomotives as well.
R. A. Lee & Co., London	1880s	Advertised as makers of 'steam engines'—nothing else known.
Leek Model Co. Ltd., Leek	1880s onwards	Brass dribbler locomotives. Later became British Modelling and Electrical Co. (q.v.)
Malins (Engineers) Ltd., Brierley Hill, Staffordshire	1936–present	See Appendix 'A'.
Meccano Ltd., Liverpool	1929–1935 1965–present	Makers of the early vertical Meccano engine. Present horizontal Meccano engine is made by Malins.
Desmond Mentha, Southport	1948	The *Rode* overtype engine.
Mersey Model Co., Liverpool	1930s 1940s	Stationary engines with some workshop accessories.
Metal Crafts Training Institute, London	1920s	'Founded to give training and employment to disabled men of H.M. Forces'. Made a P.W.E. 'mystery boat' and nine vertical oscillating-cylinder engines from 4s 6d to £5 0s 0d.
Midland Models, Birmingham	1946–1950s	*Trojan* steam plant with S.E.L. marine engine.
N.D.C.	1929–1930	Gamages' 1929 catalogue lists Power Plant No 2 by this maker. It is a good quality marine engine. Nothing else recorded.
Newton and Co., London	1860s to?	An optical company which also made dribblers and other brass engines from 1862 for some time.
Perico Models	1960s	A steam launch using Mamod's ME 1 engine.
Walter Piggott and Co., London	1920–1921	Fine range of stationary and marine engines of model calibre. Made Statham's chemical sets so possibly had taken Statham (q.v.) over.
Precision Steam Models Ltd., Selby	1970s	A reproduction dribbler loco as a kit or finished, and a simple marine engine.
L. Rees and Co., London	1948	A stationary geared oscillating-cylinder engine and a slide-valve engine.

J. & L. Randall Ltd., Potters Bar		See Signalling Equipment Ltd. and Victory Industries.
George Richardson & Co., Liverpool	c. 1860s	Locos, ships, stationary engines as well as optical and other goods.
Signalling Equipment Ltd., Potters Bar	1946–1965	The S.E.L. range of stationary engines and a die-cast marine engine.
Simpson Fawcett & Co., Leeds with Star Manufacturing Co., London	1919–1920	Good quality traditional vertical and horizontal engines with oscillating cylinders and some machine-shop accessories.
Southern Junior Aircraft Co., Brighton	c. 1947	*Southern Princess* a wooden hulled steam launch and a marine engine.
W. E. Statham Ltd., London	1860	Wide range of engines advertised in 1868. See also Walter Piggott.
Stevens Model Dockyard, London	1850s 1930s	Some early locomotives and dribblers. Later a large number were Germans in disguise.
Stuart Turner Ltd., Henley-on-Thames	1906 to present	Mainly model engines and castings. Some toy steam plants are still made.
J. W. Sutcliffe Ltd., Leeds	1920–1928	P.W.E. warships and launches.
J. Suthcliffe, Leek	— —	Became Leek Model Co. and later British Modelling and Electrical Co. (q.v.)
Technical Engineering Co., Liverpool	c. 1914	*Veitchi*—early vertical fixed-cylinder engines much like German ones.
*J. Theobald & Co., London	1880s	Novelty merchants who included some dribblers and steam turbine toys.
Tribe & Austin Ltd., Manchester later London	1919–1921	Mobile steam toys—a road wagon with gear drive, horizontal transverse boiler. Other versions as tanker, roller, tractor and two stationary engines.
Victory Industries Ltd., Surrey	c. 1948	*Miss England* P.W.E. launch.
*Warboys and Smart Ltd., Dereham	1927–1928	Stationary engines probably made for them by Bowmans.
Webb Model Fitting Company, London	c. 1950–1960	Good quality single cylinder oscillating marine engines and boilers similar to Stuart Turner's.
Whiteley Tansley & Co., Liverpool	1920	WHITANCO steam engines.
Lewis Wild & Co., London	1916	Simple vertical engines.
F. Yates and Son Ltd., London	1920s–1930s	Engines, boilers, fittings and metal hulled steam launches.

Addenda:

Glendale Engineering Co., Shipley, Yorks	c. 1950	Stationary horizontal oscillating cylinder engine.
Thomas Mayers, Brighton	c. 1960	P.W.E. boats.

Note * indicates that the company is known to have sold British engines but may not have manufactured them.

Appendix C

NOTES ON THE CARE AND RESTORATION OF EARLY TOY ENGINES

THE VERY FACT that early engines have come down to us at all from the past is surprising enough, bearing in mind all the hazards of harsh treatment by ourselves when young and the subsequent ravages of neglect and corrosion that they have suffered. Understandably enough they seldom arrive in a mint and pristine condition and the majority are at the very least in need of some loving care. Because they are pieces of miniature engineering they will usually need some engineering skills expending on them. For this reason a prospective purchaser of a collector's item should check carefully if any parts will need repair or replacement. If so and he is not able to do this himself he will have to face two related problems—one, how to find someone who will do it and, two, how to find the money to pay for it!

The basic elements of an historic (or modern) steam toy can be considered as:

a) the boiler and its fittings which will be more or less elaborate;
b) the engine, its pipework, valve gear and controls;
c) the base of a stationary engine, the hull of a boat or the frame and wheels of a locomotive, roller or traction engine.

Each of these may possibly need some attention. Boiler fittings are particularly vulnerable and are likely to need replacements whereas the engines are remarkably long suffering though if work does need to be done on them almost certainly a lathe will be required to do it. Bases and frames on the other hand will often only need gentle straightening and touching up with appropriate paints.

To consider first the boiler; this is a solid, sturdy component which is usually sound, though sometimes dented. Unfortunately there is little that can be done to remove dents, short of completely unsoldering one of the boiler ends and beating out over a wooden dowel and then re-soldering, testing etc. So, unless you are a competent amateur coppersmith avoid a boiler with dents in it where it shows. Boiler fittings are particularly subject to damage and safety-valves, whistles and steam cocks are often missing as are the wooden handles of the cocks and whistles. The chances of finding replacements for any early engines is remote, but for Mamod products the company may be able to help. Otherwise cannibalizing is the only way, apart from carefully making new ones which will often require a well-equipped workshop. The safety-valve springs will frequently have corroded away and will be heard rattling about in the boiler but small springs are easily found to replace them. The water-gauge glass tubes in the cheaper German toys are very difficult to bend and fit and sometimes it is only feasible to solder neatly the holes and put a straight dummy tube under the nickel-plated cover. On the other hand, wooden handles are easily made from dowelling, either on a small lathe or even in a hand drill with coarse and fine glass-paper, after which they should be painted dull black.

Should the boiler and its fittings have been of polished brass, as are most British engines, there should be no hesitation in cleaning them up with Brasso or the like and polishing them as nicely as possible. Those boilers however, mostly German, which were oxydized or finished in 'steel blue' should *never* be polished unless they are in so bad a state as to have lost all claim to oxydizing at all. All that should be done with the majority is to clean carefully with petrol and then rub with a cloth dipped in boiled linseed oil to give them a light protective coat.

There is, of course, always the danger that an early boiler is unsafe. However in toy engines the pressures are so low—only about one atmosphere—that even if a boiler should fail the result is unlikely to be catastrophic. Nevertheless care should be taken, and should a boiler seem to be frail it should not be steamed. A compressed air test (a car foot pump, preferably with pressure gauge, will serve) will give some useful indications of the condition of an unknown boiler and will also reveal any shortcomings of the engine as well.

There is very little that can go wrong with an oscillating-cylinder engine. Perhaps a replacement spring to hold the port faces together will be required, but this should be easy to find. With any early and neglected engine the piston should be removed and both it and the cylinder cleaned with petrol before lubricating and reassembling. The crankshaft and any gears will also repay cleaning and oiling. Sometimes a disc crank pin is bent. If so it should be carefully straightened so as to be as accurately as possible parallel to the crankshaft. This can be checked with a rule or small square.

Those engines with slide or piston-valves often have badly worn pivot pins and eccentric sheaves resulting in a lot of lost motion where this is very undesirable. A lathe is necessary to renew these and, should this prove a difficulty, serious consideration should be given to letting someone else buy the engine, particularly if it is hoped to steam it. Badly bent and/or badly soldered pipework connecting engine with boiler is a frequent and unhappy characteristic of the 'repaired' toy. It is quite incredible the botched attempts that have been made to re-solder fittings which are the result of too much solder and not enough heat. The only satisfactory way to deal with these is to re-do such joints with a good big hot iron (or a pencil thin gas flame is often better) and a good flux like Baker's Fluid. This is a very active (and hence corrosive) flux and for this reason useful on old and perhaps greasy components which are difficult to clean up. But do wash everything well with soap and hot water afterwards to get rid of any traces of the corrosive flux.

Soft solder is usually necessary for repairs to any of the hot parts — engine, boiler and steam pipes. However, for small missing or loose parts of the base (e.g. railings on mill engines) or perhaps cab or frame details, where the original lithographed or painted finish would be damaged by heat, one of the epoxy adhesives such as Araldite can often be used successfully.

The treatment of old paintwork is a subject that rouses some passion among collectors. At the one end of the scale there are those who do nothing, valuing the scratches and chippings as evidence of age and contributing to the nostalgic aura of the toy. At the other end are the complete re-painters, re-nickel platers and sometimes even re-designers as well. The vintage and veteran car people tend to be of this persuasion. What should be done depends so much upon the condition of the toy. In general there seems no doubt that as much as possible of the original should in all cases be retained. Sometimes however the paint is so decayed and the corrosion so advanced that full treatment and a re-paint must be done. Corrosion must be dealt with first by removing all loose rust with anything from a wire scratchbrush to a glass-fibre pencil eraser, depending on the extent of the damage, and then treating with a proprietory inhibitor such as Jenolite.

As far as the paintwork is concerned the first thing to do is to clean with a solvent, taking great care to remove only the dirty top layer. Then it is usually best to confine oneself to careful filling with matt diluted paints those scratched and chipped areas only, so as to preserve as much as possible of the original. Matching is a tedious business and many colours change from wet to dry, so always make sure that your chosen mixture has dried before approving it. The vast range of Humbrol paints in tiny cans make mixing a careful match easier than it was in the past, but it still takes care and time. It should be an absolute rule to leave untouched any trade marks, transfers or other indications of origin however badly damaged they may be.

It is often very difficult to decide what to do about the painted pot boilers of simple toy locomotives. They are usually scorched and if painted other than black tend to look very neglected. If a re-paint is decided upon it must be remembered that, should further steamings be attempted, the same fate will rapidly befall the new surface. On balance, attractive railway company livery notwithstanding, black is really the only practical colour for such toys if they are to be allowed to 'live' occasionally.

When all painting has been done and it is thoroughly dry a very light application with a soft cloth impregnated with linseed oil will bring up the colours, put a protective coat on and leave the engine in an attractive, cared-for condition without bright, shiny brand new colours to destroy all its period atmosphere.

Lastly, a word about storage and display. To prevent damage and corrosion — the worst enemy — it is essential to keep all metalwork dry, so boilers must always be emptied. If the engine is to be stored for a time the very best way is to seal it in a plastic bag with a few crystals of silica gel inside to absorb any residual moisture. Display is of course a purely personal thing, but it is useful to remember that, since when working the engines run at a relatively high temperature, they can be kept dry and come to no harm if set on storage heaters or above central heating radiators, where they can be admired by the collector and his guests.